SWRRB

SWRRB

A Basin Scale Simulation Model for Soil and Water Resources Management

by

J. G. Arnold, J. R. Williams, A. D. Nicks, and N. B. Sammons

TEXAS A&M UNIVERSITY PRESS
College Station
1990

Manufactured in the United States of America

FIRST EDITION

Library of Congress Cataloging-in-Publication Data

SWRRB : a basin scale simulation model for soil and water resources management / by J. G. Arnold . . . [et al.]. -- 1st ed.

p. cm.

Bibliography: p.

ISBN 0-89096-337-1

1. SWRRB (Computer program) 2. Watershed management--Computer simulation. 3. Soil management--Computer simulation. I. Arnold, J. G. (Jeffrey G.), 1959-

TC409.S87 1990

627'.01'13--dc20 89-33835

CIP

Contributors

J. G. Arnold United States Department of Agriculture
Agricultural Research Service
Grassland, Soil and Water Research Laboratory
Temple, Texas

J. R. Williams United States Department of Agriculture
Agricultural Research Service
Grassland, Soil and Water Research Laboratory
Temple, Texas

A. D. Nicks United States Department of Agriculture
Agricultural Research Service
Water Quality / Watershed Research Laboratory
Durant, Oklahoma

N. B. Sammons United States Department of Agriculture
Agricultural Research Service
Grassland, Soil and Water Research Laboratory
Temple, Texas

Contributors

J. G. Arnold — United States Department of Agriculture
Agricultural Research Service
Grassland, Soil and Water Research Laboratory
Temple, Texas

J. R. Williams — United States Department of Agriculture
Agricultural Research Service
Grassland, Soil and Water Research Laboratory
Temple, Texas

A. D. Nicks — United States Department of Agriculture
Agricultural Research Service
Water Quality and Watershed Research Laboratory
Durant, Oklahoma

Contents

Tables

Tables

SWRRB

Introduction

A model called SWRRB (Simulator for Water Resources in Rural Basins) was developed for simulating hydrologic and related processes in rural basins (Williams et al., 1985). The objective in model development was to predict the effect of management decisions on water and sediment yields with reasonable accuracy for ungaged rural basins throughout the U.S. To satisfy the objective, the model had to be (a) physically based, using readily available inputs (calibration is not possible on ungaged basins); (b) capable of computing the effects of management changes on outputs; (c) computationally efficient to allow simulation on a variety of management strategies without excessive cost; (d) capable of simulating long periods for use in frequency analysis; and (e) capable of operating on subdivided basins (soils, land use, management, etc., make subdivision necessary).

The major processes included in the model are surface runoff, percolation, return flow, evapotranspiration, transmission losses, pond and reservoir storage, sedimentation, and crop growth. The Soil Conservation Service (SCS) curve number technique (USDA, 1972) was selected for use in predicting surface runoff because (a) it is a reliable procedure that has been used for many years in the U.S.; (b) it is computationally efficient; (c) the required inputs are generally available; and (d) it relates runoff to soil type, land use, and management practices. The use of readily available daily rainfall is a particularly important attribute of the curve number technique. For many locations, rainfall data with time increments less than one day are not available. Also, daily rainfall data manipulation and runoff computation are more efficient than similar operations with shorter time increments. Traditionally, the SCS has used an antecedent rainfall index to estimate three antecedent soil moisture conditions (I-dry, II-a normal, III-wet). In reality, soil moisture varies continuously and thus curve number has many values instead of only three. Runoff prediction accuracy was increased by using a soil moisture accounting procedure (Williams and Laseur, 1976) to estimate the curve number for each storm. Although the soil moisture accounting model is superior to the antecedent rainfall method, it does not maintain a water balance and requires calibration with measured runoff data.

The CREAMS (Chemicals, Runoff, and Erosion from Agricultural Management Systems) (Knisel, 1980; Williams and Nicks, 1982) daily rainfall hydrology model overcame these deficiencies by linking the curve number technique with evapotranspiration and percolation models.

Calibration is not necessary because the CREAMS model is more physically based--the soil-water balance is related directly to curve number. Although the CREAMS daily rainfall hydrology model is more advanced than earlier curve number models, it is not applicable to complex basins. The model was developed for use on field-size areas (single land use, soil, and management practice) and does not compute water yield (return flow is neglected).

The CREAMS daily rainfall hydrology model was modified for application to large, complex, rural basins. The major changes involved were (a) the model was expanded to allow simultaneous computations on several subbasins to predict the basin water yield; (b) a return flow component was added; (c) a reservoir storage component was added for use in determining the effects of farm ponds and reservoirs on water and sediment yield; (d) a weather simulation model (rainfall, solar radiation, and temperature) was added to provide for longer term simulations and more representative weather inputs, both temporally and spatially; (e) a better method was developed for predicting the peak runoff rate; (f) a crop growth model was added to account for annual variation in growth; (g) a simple flood routing component was added; (h) components were added to simulate sediment movement through ponds, reservoirs, streams, and valleys; and (i) transmission losses were calculated. Besides water, SWRRB also simulates sediment yield from rural basins using the Modified Universal Soil Loss Equation (MUSLE) (Williams and Berndt, 1977) and a sediment routing model.

Model Description

The SWRRB model includes three major components: hydrology, weather, and sediment yield. The computer program contains 2680 FORTRAN statements in 36 subroutines. Since SWRRB operates on a daily time step, computer cost for overnight turn around is only about $0.15 per year of simulation (for a basin with four subbasins) on an AMDAHL 470 computer. The identical data set took less than one minute per year of simulation on an 80386 based microcomputer with a coprocessor. There is a special version of the model that contains a pesticide component.

Because of the provision for subdividing basins and because each subbasin can use a different rain gage, SWRRB is not limited by drainage area. Also, in the vertical direction the model is capable of working with any variation in soil properties--the soil profile is divided into a maximum of 10 layers (the top layer thickness is set at 10 mm and all other layers may have variable thickness).

Descriptions of the SWRRB components and the mathematical relationships used to simulate the processes involved follows.

A. Hydrology

The hydrology model is based on the water balance equation

$$SW_t = SW + \sum_{i=1}^{t} (R_i - Q_i - ET_i - P_i - QR_i) \quad (1)$$

where SW is the soil water content minus the 15-bar water content, t is time in days, and R, Q, ET, P, and QR are the daily amounts of precipitation, runoff, evapotranspiration, percolation, and return flow; all units are in mm.

Since the model maintains a continuous water balance, complex basins are subdivided to reflect differences in ET for various crops, soils, etc. Thus, runoff is predicted separately for each subarea and routed to obtain the total runoff for the basins. This increases accuracy and gives a much better physical description of the water balance.

1. Surface Runoff

a. Runoff Volume.--Surface runoff is predicted for daily rainfall using the SCS curve number (USDA, 1972) equation

$$Q = (R - 0.2\,s)^2 / (R + 0.8\,s) \tag{2}$$

where Q is the daily runoff, R is the daily rainfall, and s is a retention parameter. The retention parameter, s, varies among watersheds because soils, land use, management, and slope all vary. It also varies with time because of changes in soil water content. The retention parameter, s, is related to curve number (CN) using SCS equation

$$s = 254\left(\frac{100}{CN} - 1\right) \tag{3}$$

The constant, 254, in Eq. 3 gives s in mm. Thus, R and Q are also expressed in mm. The condition 2 (CN_2) or average curve number can be obtained for any area using the SCS Hydrology Handbook (USDA, 1972). Values of CN_1 and CN_3 corresponding to CN_2 are tabulated in the SCS Hydrology Handbook. For computing purposes, CN_1 and CN_3 were related to CN_2 with the equations

$$CN_1 = CN_2 - \frac{20\,(100 - CN_2)}{100 - CN_2 + \exp[2.533 - 0.063(100 - CN_2)]} \tag{4}$$

$$CN_3 = CN_2 \exp[0.00673(100 - CN_2)] \tag{5}$$

Fluctuations in soil water content cause the retention parameter to change according to the equation (Williams et al., 1984)

$$s = s_1\left(1 - \frac{SW}{SW + \exp[w_1 - w_2\,(SW)]}\right) \tag{6}$$

where SW is the soil water content in the root zone minus the wilting point (1500 kPa for many soils and crops) water content, w_1 and w_2 are shape parameters , and s_1 is the value of s associated with CN_1.

Values for w_1 and w_2 are obtained from a simultaneous solution of Eq. 6 assuming that $s = s_1$ at the wilting point and $s = s_3$ at field capacity.

$$w_1 = \ln\left(\frac{FC}{1 - \frac{s_3}{s_1}}\right) - FC + w_2\,(FC) \qquad (7)$$

$$w_2 = \frac{\ln\left(\frac{FC}{1 - \frac{s_3}{s_1}} - FC\right) - \ln\left(\frac{UL}{1 - \frac{2.54}{s_1}} - UL\right)}{UL - FC} \qquad (8)$$

where FC is the root zone water content at field capacity (33 kPa for many soils) minus the wilting point water content in mm, s_3 is the CN_3 retention parameter, and UL is the upper limit of water storage in the root zone (porosity minus wilting point) in mm. Thus, CN_1 corresponds to the wilting point and CN_3 to field capacity. At saturation, Eq. 6 allows s to approach its lower limit of 0 giving $CN \rightarrow 100$. SWRRB also includes a provision for estimating runoff from frozen soil. If the temperature in the second layer is less than 0° C, the retention parameter is reduced using the equation

$$s_f = s_1\,[1 - \exp(-0.000862\, s)\,] \qquad (9)$$

where s_f is the frozen ground retention parameter. Equation 9 increases runoff for frozen soils, but allows significant infiltration when soils are dry.

b. Peak Runoff Rate.--Peak runoff rate predictions are based on a modification of the Rational Formula.

$$q_p = (\rho)\,(r)\,(A)\,/\,360 \qquad (10)$$

where q_p is the peak runoff rate in m^3/s, ρ is a runoff coefficient expressing the watershed infiltration characteristics, r is the rainfall intensity in mm/h for the watershed's time of concentration, and A is the drainage area in ha. The runoff coefficient can be calculated for each storm if the amount of rainfall and runoff are known.

$$\rho = Q\,/\,R \qquad (11)$$

Since R is input and Q is computed by SWRRB, ρ can be calculated directly. Rainfall intensity can be expressed with the relationship

$$r = R_{tc} / t_c \tag{12}$$

where R_{tc} is the amount of rainfall in mm during the watershed's time of concentration, t_c, in h. The value of R_{tc} can be estimated by developing a relationship with total R. The Weather Service TP-40 (Hershfield, 1961) provides accumulated rainfall amounts for various durations and frequencies. Generally, R_{tc} and R_{24} (24 h duration is appropriate because SWRRB uses daily rainfall) are proportional for various frequencies. Thus,

$$R_{tc} = \alpha R_{24} \tag{13}$$

where α is a dimensionless parameter that expresses the proportion of total rainfall that occurs during t_c.

The peak runoff equation is obtained by substituting Eqs. 11, 12, and 13 into Eq. 10.

$$q_p = (\alpha)(Q)(A) / 360 (t_c) \tag{14}$$

The time of concentration can be estimated by adding the surface and channel flow times.

$$t_c = t_{cc} + t_{cs} \tag{15}$$

where t_{cc} is the time of concentration for channel flow and t_{cs} is the time of concentration for surface flow in h. The t_{cc} can be computed using the equation

$$t_{cc} = L_c / 3.6 (v_c) \tag{16}$$

where L_c is the average channel flow length for the watershed in km and v_c is the average channel velocity in m/s. The average channel flow length can be estimated using the equation

$$L_c = \sqrt{(L)(L_{ca})} \tag{17}$$

where L is the channel length from the most distant point to the subbasin outlet in km and L_{ca} is the distance along the channel to the subbasin centroid in km. Average velocity can be estimated using Manning's equation assuming a trapezoidal channel with 2:1 side slopes and a 10:1 bottom width-depth ratio. Proper substitution of these assumptions gives

$$t_{cc} = \frac{\sqrt{(L)\,(L_{ca})}\,(n)^{0.75}}{0.489\,(q_c)^{0.25}\,(\Omega)^{0.375}} \tag{18}$$

where n is Manning's "n" value, q_c is the average flow rate in m^3/s, and Ω is the average channel slope in m/m. Assuming $L_{ca} = 0.5L$ and the average flow rate is about 6.35 mm/h and is a function of the square root of drainage area yields in the final equation for t_{cc}

$$t_{cc} = \frac{0.31\,(L)\,(n)^{0.75}}{(A)^{0.125}\,(\Omega)^{0.375}} \tag{19}$$

where Ω is the average channel width in m. A similar approach is used to estimate t_{cs}.

$$t_{cs} = \lambda\,/\,3600\,(v_s) \tag{20}$$

where λ is the surface slope length in m and v_s is the surface flow velocity in m/s. Considering a strip 1 m wide down the sloping surface and applying Manning's equation gives

$$v_s = (q_s)^{0.4}\,(S)^{0.3}\,/\,(n)^{0.6} \tag{21}$$

where S is the land surface slope in m/m and q_s is the average surface flow rate. Assuming that the average flow rate is about 6.35 mm/h and making substitutions into Eqs. 20 and 21 to convert from m^3/s to mm/h and from s to h, gives the equation for estimating t_{cs}.

$$t_{cs} = (\lambda\, n)^{0.6}\,/\,18\,(S)^{0.3} \tag{22}$$

Although some of the assumptions used in developing Eqs. 19 and 22 may appear liberal, Eq. 15 generally gives satisfactory results for homogeneous watersheds. Since Eqs. 19 and 22 are based on hydraulic considerations, they are more reliable than purely empirical equations.

To properly evaluate α, variation in rainfall patterns must be considered. Some storms have short durations--most or all rain occurs during t_c causing α to approach a minimum value. All other patterns have higher α values than the uniform pattern, because R_{tc} is greater than R_{24} for all patterns except the uniform. By substituting the products of intensity and time in Eq. 13, and expression for the minimum value of α is obtained.

$$\alpha_{mn} = t_c / 24 \tag{23}$$

Thus α ranges with the limits

$$t_c / 24 \leq \alpha \leq 1.0 .$$

Although the value of α is confined between limits, there is considerable uncertainty in assigning appropriate values given only daily rainfall and simulated runoff amounts. Thus α is generated from a gamma function with the base ranging from $t_c/24$ to 1.0.

The peak of the α distribution changes for each month because of seasonal differences in rainfall intensities. The Weather Service (USDC, 1979) provides information on monthly maximum rainfall intensities that can be used to estimate the peak α for each month.

The frequency of the maximum 0.5 h rainfall amount is estimated using the equation

$$F = 1 / 2\tau \tag{24}$$

where F is the frequency of occurrence of the largest event of a total of τ events. The total number of events for each month is the product of the number of years of record and the average number of rainfall events for the month. To estimate the mean value of $\alpha_{.5}$, it is necessary to estimate the mean value of $R_{.5}$. The value of $R_{.5}$ can be computed easily if it is assumed that the maximum 0.5 h rainfall amounts are exponentially distributed. From the exponential distribution, the expression for the mean 0.5 h rainfall amount is

$$R_{.5,j} = R_{.5F,j} / -\ln F_j \tag{25}$$

where $R_{.5,j}$ is the mean maximum 0.5 h rainfall amount in mm, $R_{.5F,j}$ is the maximum 0.5 h rainfall amount for frequency F in mm, and subscript j refers to the month. The mean $\alpha_{.5}$ is computed with the equation

$$\alpha_{.5,j} = R_{.5,j} / m_j \tag{26}$$

where m is the mean amount of rainfall for each event in mm (average monthly rainfall/average number of days of rainfall) and subscript j refers to the month. Daily values of $\alpha_{.5}$ are generated from a two-parameter gamma distribution. The base of the gamma distribution is established by examining upper and lower limits of $\alpha_{.5}$. The lower limit is determined from Eq. 23. This gives a minimum of $\alpha_{.5}$ of .5/24 or 0.0208. The upper limit of $\alpha_{.5}$ is set by considering a large rainfall event. If the event is large, it is highly unlikely that all the rainfall would occur in 0.5 h ($\alpha = 1$). The upper limit of $\alpha_{.5}$ can be estimated by considering a high value for r_p (250 mm/h is generally near the upper limit of rainfall intensity).

The relation between the maximum peak rainfall rate (250 mm/h) and upper limit of $\alpha_{.5}$ can be obtained by assuming that the rainfall rate is exponentially distributed.

$$r = r_p \exp(- t / k) \tag{27}$$

where r is the rainfall rate at time t in mm/h, r_p is the peak rainfall rate in mm/h, and k is the decay constant in h. Equation 27 contains no assumption about the sequence of rainfall rates (time distribution). To compute values for r_p, Eq. 27 is integrated to give

$$R = (r_p)(k) \tag{28}$$

and

$$R_t = R\,[1 - \exp(-t / k)\,] \tag{29}$$

To determine the value of r_p, Eqs. 27 and 28 are substituted into Eq. 29 to give

$$r_p = -2\,R \ln(1 - \alpha_{.5}) \tag{30}$$

Finally, the expression for the upper limit of α is obtained by proper substitution and rearranging Eq. 30.

$$\alpha_{.5u} = 1 - \exp(-125 / R) \tag{31}$$

where $\alpha_{.5u}$ is the upper limit of $\alpha_{.5}$. The peak of the $\alpha_{.5}$ gamma distribution can be computed by differentiating the gamma function and setting the result equal to zero.

$$\alpha_{.5P,j} = \alpha_{.5,j}\,(\nu - 1) / \nu \tag{32}$$

where $\alpha_{.5P,j}$ is the $\alpha_{.5}$ value at the peak of the gamma distribution and ν is the gamma distribution shape parameter (a value of 10 is generally satisfactory).

The relationship between $\alpha_{.5}$ and α can be obtained from TP-40 (Hershfield, 1961) by fitting a log function to the 10-year frequency rainfall distribution.

$$R_t = R_6 (t / 6)^b \tag{33}$$

where b is a parameter used to fit the TP-40 relationship for any location, R_t is the rainfall amount for any time t, and R_6 is the 6-h rainfall amount in mm. The value of α is computed with the equation

$$\alpha = \alpha_{.5} (R_{tc} / R_{.5}) \tag{34}$$

2. Percolation

The percolation component of SWRRB uses a storage routing technique combined with a crack-flow model to predict flow through each soil layer. Once water percolates below the root zone, it is lost from the watershed (becomes groundwater or appears as return flow in downstream basins).

The storage routing technique is based on the equation

$$SW_i = SW_{Oi} \exp(-\Delta t / TT_i) \tag{35}$$

where SW_O and SW are the soil water contents at the beginning and end of the day in mm, Δt is the time interval (24 h), and TT is the travel time through layer i in h. Thus, the percolation can be computed by subtracting SW from SW_O.

$$O_i = SW_{Oi} [1. - \exp(-\Delta t / TT_i)] \tag{36}$$

where O is the percolation rate in mm/d.

The travel time, TT_i, is computed for each soil layer with the linear storage equation

$$TT_i = (SW_i - FC_i) / H_i \tag{37}$$

where H_i is the hydraulic conductivity in mm/h and FC is the field capacity minus wilting point water content for layer i in mm. The hydraulic conductivity is varied from the saturated conductivity value at saturation to near zero at field capacity.

$$H_i = SC_i\left(\frac{SW_i}{UL_i}\right)^{\beta_i} \tag{38}$$

where SC_i is the saturated conductivity for layer i in mm/h and β is a parameter that causes H_i to approach zero as SW_i approaches FC_i.

The equation for estimating β is

$$\beta_i = \frac{-2.655}{\log_{10}(FC_i / UL_i)} \tag{39}$$

The constant (-2.655) in Eq. 39 was set to assure $H_i = 0.002SC_i$ at field capacity.

Flow through a soil layer may be reached by a saturated lower soil layer. If the layer immediately below the layer being considered is saturated, no flow can occur regardless of the results from Eq. 38. The effect of lower layer water content is expressed in the equation

$$O_{ci} = O_i\sqrt{1 - \frac{SW_{i+1}}{UL_{i+1}}} \tag{40}$$

where O_{ci} is the percolation rate for layer i in mm/d corrected for layer i + 1 water content and O_i is the percolation computed with Eq. 36.

The crack-flow model allows percolation of infiltrated rainfall even though the soil water content is less than field capacity. When the soil is dry and cracked, infiltrated rainfall can flow through the cracks of a layer without becoming part of the layer's soil water. However, the portion that does become part of a layer's stored water cannot percolate until the storage exceeds field capacity.

Crack-flow is simulated with the relationships similar to those used to estimate percolation above field capacity. The amount of percolate caused by crack flow is estimated with the equation

$$O_i = O_{i-1} [1 - \exp(-\Delta t / TT_{ci})] \tag{41}$$

where O_{i-1} is the flow from the layer above in mm/d (R-Q for the top layer) and TT_{ci} is the crack flow travel time in h. Crack flow travel time is estimated with the equation

$$TT_{ci} = O_{i-1} / (\zeta_i) \tag{42}$$

where ζ_i is a dimensionless (0-1) soil parameter that expresses degree of cracking and is a function of soil water and clay content.

$$\zeta_i = 0.01 (CLA)_i \left(1 - \frac{SW}{SW + \exp(7.0 - 0.11\, SW)}\right) \tag{43}$$

where CLA_i is the clay content of soil layer i (0-1).

Percolation is also affected by soil temperature. If the temperature in a particular layer is 0° C or below, no percolation is allowed from that layer. Water can, however, percolate into the layer as dictated by Eqs. 36 and 41.

Since the one-day time interval is relatively low for routing flow through soils, it is desirable to divide the water into several parts for routing. This is necessary because the flow rates are dependent upon soil water content which is continuously changing. For example, if the soil is extremely wet, Eqs. 36, 37, and 38 might greatly over estimate percolation if only one routing was performed using the entire amount (SW_i-FC_i). To overcome this problem, SWRRB divides each layer's inflow into 4-mm slugs for routing. Also, by dividing the inflow into 4-mm slugs and routing each slug individually through all layers, the lower layer water content relationship (Eq. 40) is allowed to function. The lower layer water content effect cannot be calculated realistically by routing all of the input through one layer before proceeding to the next layer.

3. Lateral Subsurface Flow

Lateral subsurface flow is calculated simultaneously with percolation. Each 4-mm slug is given the opportunity to percolate first and then the remainder is subjected to the lateral flow function. Thus, lateral flow can occur when the storage in any layer exceeds field capacity after percolation. The lateral flow function (similar to Eq. 36) is expressed in the equation

$$QR_i = (SW_i - FC_i)\,[1 - \exp(-\Delta t / TT_{Ri})\,] \tag{44}$$

where QR_i is the lateral flow rate for soil layer i in mm/d and TT_{Ri} is the lateral flow travel time in d (time required for subsurface flow to travel a distance equal the land surface slope length λ).

The lateral flow travel time is estimated for each soil layer using the equation

$$TT_{RIi} = 100\left(1 - \frac{K_{Ei}}{K_{Ei} + \exp[-0.151 - 0.1625\,(K_{Ei})\,]}\right) \tag{45}$$

where K_{Ei} is the ratio of saturated hydraulic conductivity and vertical equivalent hydraulic conductivity for soil layer i. Also,

$$K_{Ei} = SC_i / K_{Vi} \tag{46}$$

where K_{Vi} is the vertical equivalent hydraulic conductivity for soil layer i in mm and SCi is the saturated conductivity of layer i in mm/h. The equation for K_V (Freeze and Cherry, 1979) is

$$K_{Vi} = d_{bi} \;/ \sum_{i=1}^{N_l} (d_i / SC_i) \tag{47}$$

where d_i is the thickness of soil layer i in mm and d_{bi} is the distance from the bottom of the soil profile to the top of soil layer i in mm. The calculation of K_{Vi} begins with the bottom soil layer and progresses upward through the profile. For the bottom layer, i = 1, and K_E is always 1. For the second layer from the bottom, i = 2, etc. This equation causes permeable layers above an impermeable layer to have short travel times. Soils without an impermeable layer will have longer travel times and thus allow more water to percolate. According to Sloan and Moore (1984), subsurface flow is likely to be significant in watersheds with soils having high hydraulic conductivities and an impermeable or semipermeable layer at shallow depths that can support a perched water table.

The lateral flow travel time is also adjusted to account for variation in soil water content and flow characteristics.

$$TT_{Ri} = (BFF)(SWC)(TT_{Rli}) \tag{48}$$

where BFF is the base flow factor and SWC is a factor for the soil water content.

$$SWC = 1 - \frac{SW}{SW + \exp[\mu_1 - \mu_2 (SW)]} \tag{49}$$

Values for μ_1 and μ_2 are obtained in a simultaneous solution of Eq. 49 assuming that $TT_{Ri} = TT_{Rli}$ at the wilting point and $TT_{Ri} = 0.1\ TT_{Rli}$ when $SW = 0.75FC$.

$$\mu_1 = 9.483 + \ln(FC_i) \tag{50}$$

$$\mu_2 = 15.958 / FC_i \tag{51}$$

The dominant mechanisms in the generation of stream flow are overland flow, saturated near-surface flow, and deep aquifer flow (Freeze, 1972; Smith and Hebbert, 1983). The baseflow factor concept was used by Boughton (1984) and ranges from zero for ephemeral streams to one for perennial streams. It allows the model to predict all of the water then eventually contributes to streamflow at the subbasin outlet. The flow is lagged using the basin flow lag time.

$$BRT = 1. - \exp(-1. / BLT) \tag{52}$$

where BLT is the basin lag time in days and BRT is a storage coefficient. Basin storage and lateral subsurface flow are calculated in three steps.

$$BS = BS_{i-1} + QR_i \tag{53}$$

$$QR_i = (BS_i)(BRT) \tag{54}$$

$$BS_i = BS_i - QR_i \tag{55}$$

where BS is the basin storage in mm and QR is the lateral subsurface flow for the entire profile in mm at day i.

4. Evapotranspiration

The evapotranspiration component of SWRRB is based on the concepts of Ritchie's ET model (Ritchie, 1972). To compute potential evaporation, the model uses the equation (Priestley and Taylor, 1972)

$$E_o = 1.28 \, (h_o) \, (\gamma) \tag{56}$$

where E_o is the potential evaporation rate in mm/d, h_o is the net solar radiation in ly, and γ is a psychometric constant. The value of h_o is calculated with the equation

$$h_o = RA \, (1 - AB) \, / \, 58.3 \tag{57}$$

where RA is the daily solar radiation in ly and AB is the albedo. The albedo is evaluated by considering the soil, crop, and snow cover. If a snow cover exists with 5 mm or greater water content, the value of albedo is set to 0.6. If the snow cover is less than 5 mm and no crop is growing, the soil albedo is the appropriate value. When crops are growing, albedo is determined using the equation

$$AB = 0.23 \, (1 - EA) + (AB_s) \, (EA) \tag{58}$$

where AB_s is the soil albedo, 0.23 is the albedo for plants, and EA is a soil cover index. The value of EA ranges from 0 to 1.0 according to the equation (Williams et al., 1984)

$$EA = \exp[-2.9 \times 10^{-5} \, (CV)\,] \tag{59}$$

where CV is the sum of above ground biomass and crop residue in kg/ha.

The value of γ in Eq. 56 can be obtained from the equation

$$\gamma = \delta_s \, / \, (\delta_s + 0.68) \tag{60}$$

where δ_s is the slope of the saturation vapor pressure curve at the mean air temperature. The expression for evaluating δ_s is

$$\delta_s = 5304 \, [\exp(21.255 - 5304 \, / \, T_K] \, / \, T_K^2 \tag{61}$$

where T_K is the daily average temperature in °K

$$T_K = T + 273 \tag{62}$$

The model computes soil and plant evaporation separately. Potential soil evaporation is predicted with the equation (Richardson and Ritchie, 1973)

$$E_{so} = E_o \exp(-0.4\ LAI) \tag{63}$$

where E_{so} is the potential evaporation rate at the soil surface in mm/d and LAI is the leaf area index defined as the area of plant leaves relative to the soil surface area. Actual soil evaporation is computed in two stages. In the first stage, soil evaporation is limited only by the energy available at the surface and thus, is equal the the potential soil evaporation. When the accumulated soil evaporation exceeds the stage one upper limit (6 mm), stage two evaporative process begins.

Stage two soil evaporation is predicted with the equation

$$E_s = 3.5\,[t^{1/2} - (t - 1)^{1/2}] \tag{64}$$

where Es is the soil evaporation rate for day t in mm/d and t is the number of days since stage two evaporation began.

The distribution of soil evaporation is determined by considering snow cover and soil water content of the top 300 mm of soil. If the water content of the snow cover is equal or greater than E_s, all of the soil evaporation comes from the snow cover. If E_s exceeds the water content of the snow cover, water is removed from the top soil layer if available. If the soil water content is reduced to the 15-bar level and E_s is not satisfied, SWRRB attempts to get water from the second layer, etc.

Plant evaporation is computed with the equations

$$E_P = (E_o)\,(LAI)\,/\,3\,, \qquad 0 \leq LAI \leq 3 \tag{65}$$

$$E_P = E_o\,, \qquad LAI > 3 \tag{66}$$

where E_P is the predicted plant evaporation rate in mm/d. If soil water is limited, plant evaporation will be reduced as described in the Plant Growth section.

5. Snow Melt

The SWRRB snow melt component is similar to that of the CREAMS model (Knisel, 1980). If snow is present, it is melted on days when the maximum temperature exceeds 0° C using the equations

$$SML = 4.57\, T_{mx}, \qquad SML < SNO \tag{67}$$

$$SML = SNO, \qquad SML \geq SNO \tag{68}$$

where SML is the snow melt rate in mm/d, T_{mx} is the daily maximum air temperature in °C, and SNO is the water content of snow before melt occurs in mm. If the initial water content of snow at the beginning of the simulation is unknown, it is estimated from the equations

$$SNO_I = R_{DEC}\,[\,(2 - T_{DEC}) / (4 - T_{DEC})\,], \qquad T_{DEC} < 2°C \tag{69}$$

$$SNO_I = 0, \qquad T_{DEC} \geq 2°C \tag{70}$$

where the subscript DEC denotes December. Melted snow is treated the same as rainfall for estimating runoff, percolation, etc.

6. Transmission Losses

Many semiarid watersheds have alluvial channels that abstract large quantities of streamflow (Lane, 1982). The abstractions, or transmission losses, reduce runoff volumes as water is lost as the flood wave travels downstream. Chapter 19 of the SCS Hydrology Handbook (USDA, 1983 update) by Leonard J. Lane describes a procedure for estimating transmission losses for ephemeral streams. This procedure is used in SWRRB when channel data (width, depth, Manning's "n") are available. It is based on the method for estimating transmission losses in the absence of observed inflow-outflow data assuming no lateral inflow or out-of-bank flow. The procedure is based on derived regression equations and enables the user to estimate transmission losses for similar channels of arbitrary length and width. Analysis of observed data resulted in regression equations of the form

$$a_r = -0.001831\,(CHK)\,DU \tag{71}$$

$$k_r = -1.09\ln(1.0 - 0.2649)\,(CHK)\,DU / P_{in} \tag{72}$$

$$b_r = e^{-k_r} \tag{73}$$

where a_r is the unit channel intercept in m^3, CHK is the effective hydraulic conductivity of the channel alluvium in mm/hr (Lane, 1982; USDA, 1983 update), DU is the duration of streamflow in h, k_r is the decay factor in m/km, P_{in} is inflow volume in m^3, and b_r is the unit channel regression slope. The regression parameters are

$$a_x = [a_r / (1 - b_r)]\,(1.0 - b_r L_c) \tag{74}$$

$$b_x = e^{-2.04\, k_r L_c w} \tag{75}$$

$$P_o = -a_x / b_x \tag{76}$$

where a_x is the regression intercept in m/km, b_x is the regression slope, w is average width of flow in m, L_c is length of reach in km, and P_o is the threshold volume for a unit channel in m^3. The prediction equation for volume after losses, Q_l, is

$$Q_l = \begin{cases} 0 & P_{in} < P_o \\ -a_x + b_x P_{in} & P_{in} > P_o \end{cases} \tag{77}$$

Transmission losses are calculated twice for each subbasin. First, losses occurring within the subbasin are estimated. The inflow volume, P is assumed to be equal the surface runoff from the subbasin. The flow duration, is estimated from

$$DU = (Q)\,(A) / (1.8\, q_p) \tag{78}$$

where Q is the surface runoff volume in mm, A is the drainage area in ha, and q_p is the peak flow rate in m^3/s. Then losses occurring from the subbasin outlet to the basin outlet are estimated. The prediction equation for peak discharge after losses q_{pl}, is

$$q_{pl} = \begin{cases} 0 & Q = 0 \\ 12.1\, a_x / DU - (1.0 - b_x)\, P_{in} + b_x q_p & Q > 0 \end{cases} \tag{79}$$

Now duration of flow from subbasin to basin outlet is estimated with Eq. 78 using Q_l and q_{pl}, and the procedure is repeated. The volume and peak rate after losses are

$$Q_l = Q - TL \tag{80}$$

where TL is the transmission loss in mm.

$$q_{pl} = Q_l A / (1.8\,DU) \tag{81}$$

Again, flow duration from subbasin to basin outlet is estimated with Eq. 78 using Q_l and q_{pl}, and the procedure is repeated.

B. Weather

The weather variables necessary for driving SWRRB are precipitation, air temperature, and solar radiation. If daily precipitation data are available, they can be input directly to SWRRB. If not, the weather component of the model can simulate daily rainfall and temperature. Solar radiation is always simulated. One set of weather variables may be simulated for the entire basin, or they can be simulated for each subbasin. Descriptions of the models used for simulating precipitation, temperature, and solar radiation follow.

1. Precipitation

The SWRRB precipitation model developed by Nicks (1974) is a first-order Markov chain model. Thus, the model must be provided as input, monthly probabilities of receiving precipitation if the previous day was dry, and monthly probabilities of receiving precipitation if the previous day was wet. If the wet-dry probabilities are not available, they can be estimated using the average number of days of precipitation,

$$P(W|D)_j = .75\, dp_j / dm_j \tag{82}$$

$$P(W|W)_j = .25 + P(W|D) \tag{83}$$

where $P(W|D)$ is the probability of rainfall after a dry day, $P(W|W)$ is the probability of rainfall after a wet day, dp is the days of precipitation in month j, and dm is the number of days in month j. Given the initial wet-dry state, the model determines stochastically if precipitation occurs. A random number (0-1) is generated and compared with the appropriate wet-dry probability. If the random number is less than or equal to the wet-dry probability, precipitation occurs on that day. Random numbers greater than the wet-dry probability give no precipitation.

When a precipitation event occurs, the amount is determined by generating from a skewed normal daily precipitation distribution.

$$X = \frac{6}{g}\left[\frac{g}{2}\left(\frac{R - m}{\sigma}\right)^{1/3}\right] + \frac{g}{6} \tag{84}$$

where X is the generated normal deviate, R is the daily rainfall amount in mm, and m, σ, and g are the mean, standard deviation, and skew coefficient for daily rainfall for a particular month.

If precipitation is to be simulated for each subbasin, the amount computed using Eq. 84 is assumed to be the mean for all gages for the day. The center of the storm is located in a rectangle with boundaries set at a distance of 100 km from the basins' maximum and minimum x and y rain gage coordinates. Thus, the storm center could be located in the basin or as much as 100 km in any direction from the basin. Each storm center is defined by drawing two random numbers--one for the x scale and one for the y scale. Rainfall at each gage is computed using an area reduction function (Nicks and Igo, 1980)

$$RF = \frac{1 - (AO)\,(DUR)^{-0.148}}{337.5 + 1.09\,AO} \tag{85}$$

where RF is a reduction ratio (0-1), DUR is the rainfall duration in h, and AO is area represented by the rain gage in km^2. The circular area, AO, is calculated using the distance between the gage and the storm center to compute the diameter.

$$AO = 0.7854\,(x_c - x_i)^2 + (y_c - y_i)^2 \tag{86}$$

where x and y are the coordinates of the rain gages in km and subscripts c and i refer to the storm center and the gage for which RF is being computed. Since only daily rainfall is considered, the duration is unknown. Thus, it must be estimated using monthly mean durations. Limited experiments have shown that rainfall duration is exponentially distributed. Therefore, the daily duration is generated from an exponential distribution using the monthly mean duration.

$$DUR = -\ln(\chi)\,\overline{DUR_j} \tag{87}$$

where χ is a uniform random number and $\overline{DUR_j}$ is the mean rainfall duration for individual events for month j.

Finally, rainfall is generated for each gage using the equation

$$R_i = \frac{(RF_i)\,(R)\,(N)}{\sum_{i=1}^{N} RF_i} \tag{88}$$

where R_i is the daily rainfall for rain gage i, RF_i is the reduction factor, R is the mean daily rainfall generated from Eq. 84, and N is the number of subbasins.

The amount of daily precipitation is partitioned between rainfall and snowfall using average daily air temperature. If the average daily air temperature is 0° C or below, the precipitation is snowfall--otherwise it is rainfall.

2. Air Temperature

SWRRB generates daily maximum air temperature using the normal distribution equations (Nicks and Harp, 1980)

$$T_{mx} = \bar{T}_{mx,j} + (U_{Tmx,j})\,(\chi_{Tmx})\,(B_F) \tag{89}$$

and

$$T_{mn} = \bar{T}_{mn,j} + (U_{Tmn,j})\,(\chi_{Tmn})\,(B_F) \tag{90}$$

where T_{mx} and T_{mn} are the generated maximum and minimum temperatures in °C, $\bar{T}_{mx}$ and $\bar{T}_{mn}$ are the daily mean maximum and minimum temperatures in month j in °C, $U_{Tmx,j}$ and $U_{Tmn,j}$ are the standard deviations of daily maximum and minimum temperatures for month j in °C, χ_{Tmx} and χ_{Tmn} are uniform random numbers drawn for T_{mx} and T_{mn}, and B_F is a weighting factor based on wet-dry rainfall probabilities. The weighting factors are used to provide more deviation in temperatures when the weather changes and for rainy days. Values of B_F for each of the four possible wet-dry situations are given by the equations

$$B_F\,(W|D) = [1 - P(W|D)\,]\,/\,PF \tag{91}$$

$$B_F\,(W|W) = [1 - P(W|W)\,]\,/\,PF \tag{92}$$

$$B_F\,(D|D) = P(W|D)\,/\,PF \tag{93}$$

$$B_F\,(D|W) = P(W|W)\,/\,PF \tag{94}$$

where P(W|D) is the probability of rainfall occurring after a dry day, P(W|W) is the probability of rainfall after a wet day, and PF is a probability factor used to assure proper standard deviations of generated data. The equation for computing PF is

$$PF = P(W|D)\,[1 - P(W|D)\,] + P(W|W)\,[1 - P(W|W)\,] \tag{95}$$

Equation 95 assures that

$$\sum_{i=1}^{4} (B_{Fi})\,(P_i) = 1$$

and thus the average value of B_F in Eqs. 89 and 90 is 1.

3. Solar Radiation

Solar radiation is also generated from the normal distribution using the equation (Nicks and Harp, 1980)

$$RA_{IDA} = \overline{RA}_j + (U_{RA,j})\,(\chi_{RA})\,(B_F) \tag{96}$$

where RA_{IDA} is the generated solar radiation for day IDA in Ly, $\overline{RA}_j$ is the mean daily radiation for month j in Ly, U_{RA} is the standard deviation for daily radiation for month j in Ly, and χ_{RA} is a uniform random number drawn for RA. Solar radiation is estimated with the equation

$$RA_{mx,IDA} = \frac{711}{D^2}\left(\phi \sin(\frac{2\pi}{360}LAT)\sin(\theta) + \cos(\frac{2\pi}{360}LAT)\cos(\theta)\sin(\phi)\right) \tag{97}$$

where $RA_{mx,IDA}$ is the maximum solar radiation for day IDA in Ly, D is the earth's radius vector, ϕ is the sun's half day length in radians, LAT is the basin's latitude, and θ is the sun's declination angle in radians. Earth's radius vector can be computed for any day (IDA) using the equation

$$D = \sqrt{\frac{1.0}{1.0 + 0.0335\sin[\frac{2\pi}{365}(IDA + 88.2)\,]}} \tag{98}$$

The sun's declination angle is computed with the equation

$$\theta = 0.4102 \sin[\frac{2\pi}{365}(IDA - 80.25)] \tag{99}$$

Finally, the sun's half day length is compared with the relationships

$$\phi = \cos^{-1}[\tan(\frac{2\pi}{360} LAT)\tan(\theta)], \qquad -1 \le \theta \le 1 \tag{100}$$

$$\phi = 0, \qquad \theta > 1 \tag{101}$$

$$\phi = \pi, \qquad \theta < -1 \tag{102}$$

The standard deviation of solar radiation, U_{RA}, can be estimated using Eq. 97 and the mean radiation, RA.

$$U_{RA,j} = (RA_{mx,IDA} - \overline{RA_j}) / 4 \tag{103}$$

The value of RA_{mx} is computed for a day in the middle of each month. Eq. 103 assumes that the difference between RA_{mx} and $\overline{RA}$ is four standard deviations.

C. Flood Routing

Since SWRRB is primarily a long-term water and sediment yield simulator, a high degree of accuracy in predicting hydrographs is not as necessary as for other applications like flood control planning and flood forecasting. Also, SWRRB must operate as efficiently as possible to be useful in water resources planning that requires long-term simulations of numerous management strategies. Therefore, flood routing is not performed in SWRRB and daily basin outflow is estimated by summing subbasin outflows. All surface runoff reaches the basin outlets on the day of the event. Subsurface flow is lagged as a function of basin lag time.

Basin peak flow rates are estimated using the modified Rational equation as described in Section A.1.b. The peak rates are determined for use in developing a peak-flow-frequency distribution. The flow-frequency distribution is useful in checking simulated results against measured values from the basin or from similar basins. Accurate prediction of subbasin peak rates is also important since they are used in estimating subbasin sediment yields. Comparison of measured and SWRRB-simulated peak rates are shown in the Model Evaluation section.

D. Ponds and Reservoirs

1. Ponds

This component of SWRRB was designed to account for the effects of farm ponds on water yield. The water balance equation is

$$VM = VM_O + QI - QO - EV - SEP \tag{104}$$

where VM_O is the volume of water stored in all ponds within a subbasin at the beginning of the day, VM is the volume at the end of the day, QI is the inflow during the day, QO is the outflow, EV is the evaporation, SEP is the seepage, and all units are m^3. The inflow, QI is composed of surface runoff from the total pond drainage area and rainfall on the water surface area. Outflow occurs when the volume exceeds the permanent pool storage capacity and is described with the equation

$$QO = VM - VM_{mx}, \qquad VM > VM_{mx} \tag{105}$$

$$QO = 0, \qquad VM \le VM_{mx}$$

where VM_{mx} is the maximum permanent pool storage of all ponds in the subbasin in m^3.

The evaporation is computed with the equation

$$EV = 10\,(\eta)\,(E_o)\,(SA) \tag{106}$$

where η is an evaporation coefficient (0.6) and SA is the surface area of all ponds in the subbasin in ha.

Seepage from the ponds is computed with the equation

$$SEP = 240\,(SC)\,(SA) \tag{107}$$

where SC is the saturated conductivity of the pond bed in mm/h.

Since pond surface area is required for computing evaporation and seepage, a relationship between surface area and volume is necessary. Data from a large number of stock ponds and small reservoirs in Texas and Oklahoma (USDA, 1957) indicate that surface area can be calculated with the equation

$$SA = \omega_1 (VM)^{\omega_2} \tag{108}$$

where ω_1 is a parameter (1.3×10^{-4}) and ω_2 is a fairly constant parameter (0.9). The SWRRB assumes $\omega_2 = 0.9$ and determines ω_1 for each subbasin using given SA_{mx} and VM_{mx}.

2. Reservoirs

Although this component was mainly designed to simulate flow through small reservoirs like those constructed on SCS PL566 projects, it can also be used on larger reservoirs. The reservoir water balance component is similar to the pond component except it allows flow from principal and emergency spillways. The reservoir outflow function is expressed in the equation

$$QOR = VR - VR_F, \qquad VR > VR_F \tag{109}$$

$$QOR = (RR)(\Delta t), \qquad VR_S < VR \leq VR_F$$

$$QOR = 0. \qquad VR < VR_S$$

where QOR is the daily outflow in m^3, VR is the volume of water in the reservoir in m^3, VR_F is the reservoir capacity at the emergency spillway crest in m^3, RR is the principal spillway release rate in m^3/s, and VR_S is the reservoir capacity at the principal spillway crest in m^3.

The relationships (Eqs. 106, 107, and 108) used to estimate evaporation and seepage from ponds are also applicable to reservoirs. However, the method for estimating ω_1 and ω_2 is slightly different. Since the surface areas and volumes for the principal and emergency spillway crest elevations are generally readily available, those values can be used for a simultaneous solution of Eq. 108. The resulting equations are

$$\omega_1 = (\log SA_F - \log SA_S) / (\log VR_F - \log VR_S) \tag{110}$$

$$\omega_2 = SA_F / VR^{\omega_2} \tag{111}$$

where SA is the reservoir surface area and subscripts F and S refer to emergency and principal spillway crest elevations, respectively.

E. Sediment Yield

Sediment yield is computed for each subbasin with the Modified Universal Soil Loss Equation (MUSLE) (Williams and Berndt, 1977).

$$Y = 11.8\,(V\ q_p)^{0.56}\,(K)\,(C)\,(PE)\,(LS) \tag{112}$$

where Y is the sediment yield from the subbasin in t, V is the surface runoff volume for the subbasin in m^3, q_p is the peak flow rate for the subbasin in m^3/s, K is the soil erodibility factor, C is the crop management factor, PE is the erosion control practice factor, and LS is the slope length and steepness factor.

The LS factor is computed with the equation (Wischmeier and Smith, 1978)

$$LS = (\lambda / 22.1)^{\xi}\,(65.41\,S^2 + 4.565\,S + .065) \tag{113}$$

The exponent ξ varies with slope and is computed in SWRRB with the equation

$$\xi = 0.6\,[1 - \exp(-35.835\,S)\,] \tag{114}$$

The crop management factor, C, is evaluated for all days when runoff occurs using the equation

$$C = \exp[\,(-0.2231 - CVM)\,\exp(-0.00115\,CV) + CVM] \tag{115}$$

where CV is the soil cover (above ground biomass + residue) in kg/ha and CVM is the minimum value of C. The value of CVM is estimated from the average annual C factor using the equation

$$CVM = 1.463\,\ln(CVA) + 0.1034) \tag{116}$$

The value of CVA for each crop is determined from tables prepared by Wischmeier and Smith (1978). Values of K are contained in the SCS Soils-5 data base, and PE factors can be estimated for each subbasin using information contained in Technical Release No. 51 (USDA, 1975).

F. Sediment Routing

1. Ponds and Reservoirs

Inflow sediment yield to ponds and reservoirs (P/R) is computed with MUSLE as described in the Sediment Yield section. The outflow from P/R is calculated with the equation

$$Y = (c_o)(QO), \qquad QO > 0 \qquad (117)$$

$$Y = 0, \qquad QO = 0$$

where Y is the P/R outflow sediment yield in t, c_o is the outflow sediment concentration in t/m^3, and QO is the outflow in m^3. The outflow concentration is a function of the P/R concentration at the beginning and end of the day.

$$c_o = (cs_1 + cs_2)/2 \qquad (118)$$

where cs_1 and cs_2 are the P/R concentrations at the beginning and end of a day in t/m^3.

The initial P/R concentration is input to SWRRB. The inflow concentration can be calculated since Y and Q are simulated, but the final P/R concentration is unknown. It can be computed using the continuity equation

$$VM_2\, cs_2 = VM_1\, cs_1 + QI\, c_I - QO\, c_o \qquad (119)$$

where VM_1 and VM_2 are the P/R storage volumes at the beginning and end of a day in m^3, QI is the inflow volume in m^3, and c_I is the inflow concentration in t/m^3. Substituting Eq. 118 into Eq. 119 and rearranging yields and expression for the final concentration

$$cs_2 = \frac{VM_i\, cs_i + QI\, c_I - (QO/2)\, cs_1}{VM_2 + (QO/2)} \qquad (120)$$

Between storms the final P/R concentration decreases to an equilibrium concentration according to the equation

$$cs = (cs_2 - cs_e)\exp(-k_s\, t\, D50) + cs_e \qquad (121)$$

where cs is the P/R concentration t days after the value of cs_2 is obtained, k_s is the decay constant, D50 is the median particle size of the inflow sediment in μm, and c_{Se} is the equilibrium P/R sediment concentration (input to SWRRB). A value of k_s is evaluated by assuming that 99% of the 1 μm particles are settled within 25 days ($k_s = 0.184$).

2. Channel and Floodplain

The sediment routing model consists of two components operating simultaneously (deposition and degradation). The deposition component is based on fall velocity, and the degradation component is based on Bagnold's (1977) stream power concept (Williams, 1980).

The sediment size distribution of the detached sediment is estimated from the primary particle size distribution (Foster et al., 1980). The values are typical of many midwestern soils.

$$PSA = 2.49\,(SAN)\,(1. - CLA) \tag{122}$$

$$PSI = 0.13\,SIL \tag{123}$$

$$PCL = 0.20\,CLA \tag{124}$$

$$SAG = \begin{cases} 2.0\,CLA & \text{for } CLA < 0.25 \\ 0.28\,(CLA - 0.25) + 0.5 & \text{for 0.25 8P10Mâ } CLA \le 0.5 \\ 0.57 & \text{for } CLA > 0.5 \end{cases} \tag{125}$$

$$LAG = 1.0 - PSA - PSI - PCL - SAG \tag{126}$$

where SAN, SIL, and CLA are the fractions of primary sand, silt, and clay in the original soil mass, and PSA, PSI, PCL, SAG, and LAG are the fractions of sand, silt, clay, small aggregates, and large aggregates for the detached sediment before deposition. The corresponding particle diameters are

$$\begin{aligned} PSA &= 0.20 \text{ mm} \\ PSI &= 0.01 \text{ mm} \\ PCL &= 0.002 \text{ mm} \\ SAG &= 0.03 \text{ mm} \\ LAG &= 0.50 \text{ mm} \end{aligned} \tag{127}$$

The particle size distribution of the sediment reaching the subbasin outlet is estimated using a routing coefficient technique (Williams, 1980). First, the delivery ratio is estimated for each runoff event using the equation

$$DR_{SB} = (q_p / r_{ep})^{0.56} \qquad (128)$$

where DR_{SB} is the sediment delivery ratio (sediment yield divided by gross sheet erosion), q_p is the peak runoff rate in mm/h, and r_{ep} is the peak rainfall excess rate in mm/h. Eq. 128 is based on sediment yield estimates using MUSLE (Williams and Berndt, 1977). The rainfall excess rate cannot be evaluated directly because the hydrology model only predicts the total daily runoff volume. An estimate of the rate can be obtained, however, using the equation

$$r_{ep} = r_p - f \qquad (129)$$

where r_p is the peak rainfall rate in mm/h and f is the average infiltration rate in mm/h. The average infiltration rate can be computed from the equation

$$f = (R - Q) / DUR \qquad (130)$$

where DUR is the rainfall duration in h. The rainfall duration can be estimated by solving Eqs. 28 and 29 for t when $R_t/R = 0.99$.

$$DUR = 4.605\, R / r_p \qquad (131)$$

The routing coefficient, β_r, is estimated with the equation (Williams, 1980)

$$\beta_r = -\ln (DR_{SB}) / 4.47 \qquad (132)$$

The particle size distribution of the sediment reaching the subbasin outlets is computed with the following equations

$$PSZ_j = PCT_j\, e^{-\beta_r \sqrt{dj}} \qquad (133)$$

$$TOT = \sum_{j=1}^{NSZ} PSZ_j \qquad (134)$$

$$PSZ_j = PSZ_j / TOT \qquad (135)$$

where PSZ_j is the fraction of particle size j in the sediment reaching the subbasin outlet, PCT_j is the fraction of particle size j in the detached sediment within the subbasin, and d_j is diameter of particle size j in μm.

Deposition in the stream channel from subbasin to basin outlet is based on the fall velocity of the sediment particles. With a temperature of 22°C and a sediment density of 1.2 t/m^3, Stokes' Law for fall velocity becomes

$$v_j = 411\, d_j^2 \tag{136}$$

where v_j is the fall velocity in m/h, and d_j is the sediment particle diameter in mm for particle size j. The depth (y_j) that sediment of particle size d_j will fall during time, TT, is

$$y_j = v_j\, TT \tag{137}$$

The travel time is calculated by first determining the flow depth, y_c. The flow depth is calculated with Manning's equation for rectangular channel and assuming $w >> y_c$.

$$y_c = (q_p\, n / w\, S_c^{\,.5})^{.6} \tag{138}$$

where w is the channel width in m, S_c is the channel slope in m/m, and y_c is the flow depth in m. The velocity in the channel (assuming a rectangular channel) is

$$v_c = q_p / (w\, y_c) \tag{139}$$

and travel time is

$$TT = L_c / (3.6\, v_c) \tag{140}$$

where L_c is the channel length in km, v_c is the velocity in m/s, and TT is the travel time in h.

The sediment delivery ratio (DR) through the reach is estimated for each particle size with the equations

$$DR = 1 - 0.5\, v_j\, TT / y_c \qquad \text{If } y_j < y_c \tag{141}$$

$$DR = 0.5\, y_c / v_j\, TT \qquad \text{If } y_j > y_c \tag{142}$$

Finally, deposition is calculated by

$$DEP = YD_j\, (1 - DR) \tag{143}$$

where YD_j is the sediment of particle size j entering the channel in t.

Stream power is used to predict degradation in the routing reaches. Williams (1980) used Bagnold's (1977) definition of stream power to develop a method for determining degradation in channels. Bagnold defined stream power, SP, with the equation

$$SP = \gamma_w q S_w \tag{144}$$

where γ_w is the density of the flowing water in t/m^3, q is the flow rate in m^3/s, and S_w is the water surface slope in m/m. By applying stream power to bed load predictions (Bagnold, 1977) and estimating model parameters (Williams, 1980), the equation for sediment reentrained, DEG_R, is

$$DEG_R = \alpha_{sp} \gamma_w^{1.5} \Delta t \sum_{t=0}^{DU} (q_t S_{wt})^{1.5} \tag{145}$$

where α_{sp} is a parameter dependent on maximum stream power and q is flow rate in m^3/s. Since the model only computes total sediment yields and not sediment graphs, only one time increment is used in Eq. 145 resulting in

$$DEG_R = \alpha_{sp} \gamma_w{}^{1.5} DU\, w\, (PR_f\, d_w\, S_w\;\; v_c)^{1.5} \tag{146}$$

where d_w is the flow depth in m, and PR_f is a peak rate adjustment factor.

PR_f is allowed to vary by event by assuming a three-part triangular hydrograph with shape assumptions shown in the figure below.

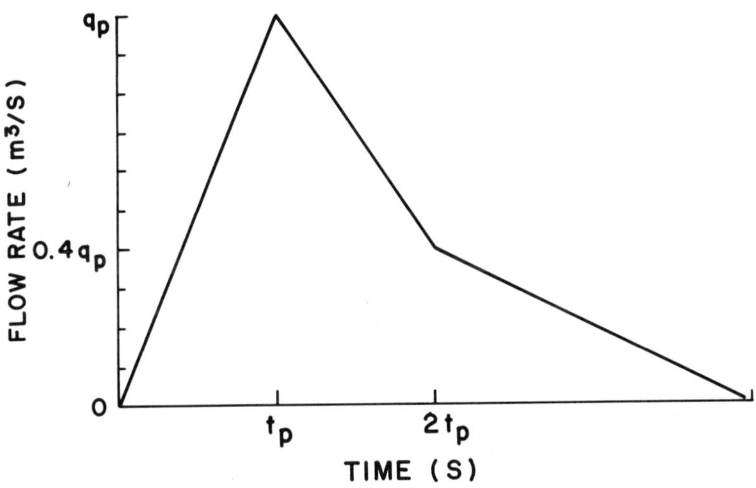

By setting Eqs. 145 and 146 equal and solving for PR_f for several hypothetical hydrographs with B_t varying from 0.05 to 0.5, a linear relationship between B_t and PR_f was found

$$PR_f = 0.5\, B_t + 0.234 \tag{147}$$

where B_t is the fraction of the hydrograph in the first part of the triangle and

$$B_t = 0.5\, t_p\, q_p / V \tag{148}$$

where t_p is the hydrograph time to peak in s (t_p is assumed to equal t_c).

The parameter α can be estimated with the equation

$$\alpha_{sp} = (\gamma_w\, q\, S_c)_{mx}^{-0.5} \tag{149}$$

where the subscript mx refers to the maximum flow expected in the routing reach for extreme events. The value of q is assumed to equal some maximum rainfall intensity (250 mm/hr) and α_{sp} becomes

$$\alpha_{sp} = (69.44\, \gamma_w\, DA\, S_c)^{-0.5} \tag{150}$$

All of the stream power is used for reentrainment of loose and deposited material until all of the material has been removed. When this occurs, degradation of the bed material, DEG_B, begins and is calculated by

$$DEG_B = K\, C\, DEG_R \tag{151}$$

where K and C are MUSLE factors for the stream channel. Total degradation, DEG, is the sum of the reentrainment and bed degradation components. This amount is also allowed to be redeposited before reaching the basin outlet.

$$DEG = (DEG_R + DEG_B)\,(1 - DR) \tag{152}$$

Finally, the amount of sediment reaching the basin outlet, SED_T, is

$$SED_T = Y - DEP + DEG_R + DEG_B \tag{153}$$

G. Soil Temperature

Daily average soil temperature is simulated at the center of each soil layer. The basic soil temperature equation is

$$T(Z,t) = \bar{T} + \frac{AM}{2} \exp(-Z/DD) \cos[\frac{2\pi}{365}(t-200) - Z/DD] \quad (154)$$

where Z is depth from the soil surface in mm, t is time in d, $\bar{T}$ is the average annual air temperature in °C, AM is the annual amplitude in daily average temperature in °C, and DD is the damping depth for the soil in mm. Equation 154 provides estimates of air temperature (Z=0) as well as soil temperature. Since air temperature is provided by the weather component of SWRRB, the soil temperature model should be capable of using these air temperatures as drivers. Otherwise, Eq. 154 would predict the same temperatures for a given day each year. To allow simulated air temperature to be used as the soil temperature driver, equations were developed to estimate soil surface temperature.

$$TGB_{IDA} = (z_j)(T_{mx}) + (1 - z_j)(T_{mn}), \qquad R = 0 \quad (155)$$

$$TGB_{IDA} = T_{mn}, \qquad R > 0 \quad (156)$$

where TGB_{IDA} is the bare soil surface temperature in °C, T_{mx} and T_{mn} are the maximum and minimum daily air temperature in °C, z is a weighting factor for estimating soil surface temperature on dry days, and j specifies the month. Obviously, Eq. 156 uses the minimum air temperature to estimate surface temperature on rainy days. Higher temperatures are estimated on dry days using Eq. 155. The value of z is determined by considering the average T_{mx}, T_{mn}, and number of rainy days in this month.

$$z_j = \frac{\dfrac{(T_{mxj} + T_{mnj}/2)(ND_j) - (T_{mnj})(NRD_j)}{NDD_j} - T_{mnj}}{T_{mxj} - T_{mnj}} \quad (157)$$

where ND is the number of days in a month, NRD is the number of rainy days, and NDD is the number of dry days. Equation 157 assures that the long-term average soil surface temperature and air temperature are equal.

The soil surface temperature is also affected by residue and snow cover. This effect can be simulated by lagging the predicted bare surface temperature using the equation

$$TG_{IDA} = (bcv)(TGB_{IDA-1}) + (1 - bcv)(TGB_{IDA}) \tag{158}$$

where bcv is a lagging factor for simulating residue and snow cover effects on surface temperature. The value of bcv is 0 for bare soil and approaches 1.0 as cover increased as expressed in the equation

$$bcv = \max \begin{cases} \dfrac{CV}{CV + \exp(7.563 - 1.297 \times 10^{-4}\, CV)} \\ \dfrac{SNO}{SNO + \exp(6.055 - 0.3022\, SNO)} \end{cases} \tag{159}$$

where CV is the sum of above ground biomass and crop residue in kg/ha and SNO is the water content of the snow cover in mm.

The soil temperature at any depth is estimated with Eq. 154 by substituting TG for T(0,t). TG is a better estimate of the surface temperature than T(0,t) because current weather and cover conditions are considered. At soil surface (Z = 0), the proper substitution can be accomplished by adding TG and subtracting T(0,t) from Eq. 154. Differences between TG and T(0,t) are damped as Z increases. Thus, the final equation for estimating soil temperature at any depth is

$$T(Z,t) = T + [(\frac{AM}{2} + TG - T(0,t)]\cos[\frac{2\pi}{365}(t - 200) - \frac{Z}{DD})]\exp(\frac{-Z}{DD}) \tag{160}$$

The damping depth (DP) is a function of soil bulk density and water content as expressed in the equations

$$DP = 1000 + \left(2500\, BD / [BD + 686 \exp(-5.63\, BD)]\right) \tag{161}$$

$$S_P = SW / [(0.356 - 0.144\, BD)\, ZM] \tag{162}$$

$$DD = DP \exp\left(\ln(500 / DP)\, [(1 - S_P) / (1 + S_P)]^2\right) \tag{163}$$

where DP is the maximum damping depth for the soil in mm, BD is the soil bulk density in t/m^3, ZM is the distance from the bottom of the lowest soil layer to the surface in mm, and S_P is a scaling parameter.

H. Crop Growth

The crop model in SWRRB is a simplification of the EPIC crop model (Williams et al., 1984). SWRRB uses EPIC concepts of phenological crop development based on daily accumulated heat units, harvest index for partitioning grain yield, Monteith's approach (Monteith, 1977) for potential biomass, and water and temperature stress adjustments. However, the detailed EPIC root growth and nutrient cycling models are not included. A single model is used for simulating all the crops considered. SWRRB is capable of simulating crop growth for both annual and perennial plants. Annual crops grow from planting date to harvest date or until the accumulated heat units equal the potential heat units for the crop. Perennial crops maintain their root systems throughout the year, although the plant may become dormant after frost.

Phenological development of the crop is based on daily heat unit accumulation. It is computed using the equation

$$HU_i = (T_{mx,i} + T_{mn,i} / 2) - T_{b,j}, \qquad HU_k \geq 0 \tag{164}$$

where HU, T_{mx}, and T_{mn} are the values of heat units, maximum temperature, and minimum temperature in °C on day i, and T_b is the crop-specific base temperature in °C (no growth occurs at or below T_b) of crop j. A heat unit index (HUI) ranging from 0 at planting to 1 at physiological maturity is computed as follows.

$$HUI_i = \frac{\sum_{k=1}^{i} HU_k}{PHU_j} \tag{165}$$

where HUI is the heat unit index for day i and PHU is the potential heat units required for maturity of crop j. The value of PHU is calculated by the model from normal planting and harvest dates.

1. Potential Growth

Interception of solar radiation is estimated with Beer's law equation (Monsi and Saeki, 1953)

$$PAR_i = 0.02092\,(RA)_i\,[1. - \exp(-0.65\,LAI)\,]_i \tag{166}$$

where PAR is photosynthetic active radiation in MJ/m^2, RA is solar radiation in Ly, LAI is the leaf area index, and subscript i is the day of the year. Using Monteith's approach (Monteith, 1977), potential increase in biomass for a day can be estimated with the equation

$$\Delta B_{p,i} = (BE)_j \, (PAR)_i \tag{167}$$

where ΔB_p is the daily potential increase in total biomass in kg/ha and BE is the crop parameter for converting energy to biomass in kg m^2/ha MJ.

LAI is simulated as a function of heat units and biomass. LAI is estimated with the equations

$$LAI_i = \frac{(LAI_{mx})\,(B_{AG})}{[B_{AG} + \exp(9.5 - .0006\, B_{AG})\,]}, \quad HIU_i \leq DLAI \tag{168}$$

$$LAI_i = (16)\,(LAI_{mx})\,(1 - HIU_i)^2, \quad HIU_i > DLAI \tag{169}$$

where LAI_{mx} is the maximum LAI potential for the crop, B_{AG} is aboveground biomass in kg/ha, and DLAI is the fraction of the growing season when LAI starts declining ($\approx$.75).

The fraction of total biomass partitioned to the root system normally decreases from 0.3 to 0.5 in the seedling to 0.05 to 0.20 at maturity (Jones, 1985). The model estimates the root fraction to range linearly from 0.4 at emergence to 0.2 at maturity. Thus, the daily root fraction is computed with the equation

$$RWT_i = (0.4 - 0.2\, HUI_i) \tag{170}$$

where RWT is the fraction of total biomass partitioned to the root system on day i. Thus, B_{AG} is calculated from the equation

$$B_{AG} = (1 - RWT_i)\,(B_{TOTi}) \tag{171}$$

where B_{TOT} is total biomass in kg/ha on day i.

2. Crop Yield

The economic yield of most grain, pulse, and tuber crops is a reproductive organ. Crops have a variety of mechanisms which insure that their production is neither too great to be supported by the vegetative components nor too small to insure survival of the species. As a result, harvest index (economic yield/above-ground biomass) is often a relatively stable value across a range of environmental conditions. Crop yield is estimated using the harvest index concept.

$$YLD_j = (HI_j)(B_{AG}) \tag{172}$$

where YLD is the amount of the crop removed from the field in kg/ha, HI is the harvest index at harvest, and B_{AG} is the above-ground biomass in kg/ha for crop j. Harvest index increases non-linearly from zero at planting using the equation

$$HIA_i = HIO_j (HUFH_i - HUFH_{i-1}) \tag{173}$$

where HIA is the harvest index on day i, HIO is the harvest index under favorable growing conditions, and HUFH is the heat unit factor that affects harvest index on day i and on the previous day i-1.

The harvest index heat unit is computed with the equation

$$HUFH_i = \frac{HUI_i}{HUI_i + \exp(6.50 - 10.0\, HUI_i)} \tag{174}$$

The constants in Eq. 174 are set to allow $HUFH_i$ to increase from 0.1 at $HUI_i = 0.5$ to 0.92 at $HUI_i = 0.9$. This is consistent with economic yield development of grain crops which produce most economic yield in the second half of the growing season.

Most grain crops are particularly sensitive to water stress from shortly before until shortly after anthesis, when major yield components are determined (Doorenbos and Kassam, 1979). Optimum conditions for growth may reduce harvest index slightly if dry matter accumulation is large and economic yield is limited by sink size. The harvest index is affected by water stress using the equation

$$HIA_i = \frac{HIA_i}{1. + (WSYF_j)(FHU_i)(0.9 - WS_i)} \tag{175}$$

where HIA is the adjusted harvest index, WSYF is a crop parameter expressing drought sensitivity, FHU is a function of crop stage, and WS is the water stress factor for day i. Notice that harvest index may increase slightly on days with WS values greater than 0.9. The crop stage factor, FHU, is estimated with the equation

$$FHU_i = \sin\left(\frac{\pi}{2}\left(\frac{HUI_i - 0.3}{0.3}\right)\right), \quad 0.3 \leq HUI_i \leq 0.9 \tag{176}$$

$$FHU_i = 0., \quad HUI < 0.3 \quad \text{or} \quad HUI > 0.9$$

Thus, water stress only affects harvest index between 0.3 and 0.9 of maturity with the greatest effect occurring at 0.6.

3. Growth Constraints

The potential biomass predicted with Eq. 167 is adjusted daily if one of the plant stress factors is less than 1.0 using the equation

$$\Delta B = (\Delta B_p)(REG) \tag{177}$$

where REG is the one crop growth regulating factor (the minimum stress factor).

a. Water Stress.--The water stress factor is computed by considering supply and demand in the equation

$$WS = \frac{\sum_{i=1}^{N_l} u_i}{E_P} \tag{178}$$

where WS is the water stress factor (0-1) and u_i is plant water use in layer i in mm. The value of E_P is predicted in the evapotranspiration model.

A model (Williams and Hann, 1978) for simulating plant water uptake is used to estimate the E_P distribution. Root depth is calculated with the equation

$$RD_n = 2.5 \frac{\sum_{i=1}^{n} E_{Pi}}{\sum_{i=1}^{N_g} E_{Pi}} \tag{179}$$

$$0 \leq RD_n \leq 1.0$$

where RD_n is the fraction of the root zone that contains roots on day n and N_g is the number of days in the growing season.

The potential water use in each soil layer is estimated with the equation

$$u_{pi} = \frac{E_P}{1 - \exp(-\Lambda)} \left(1 - \exp[-\Lambda\,(\frac{RD_i}{RZ})\,]\right) \sum_{j=1}^{i-1} u_j \tag{180}$$

where u_{pi} is the potential water use rate from layer i in mm/d, RZ is the root zone depth in mm, and Λ is the use rate-depth parameter. The details of evaluating Λ are given by Williams and Hann (1978). The value of Λ used in SWRRB (3.065) was determined assuming that about 30% of the total water use comes from the top 10% of the root zone. Equation 180 allows roots to compensate for water deficits in certain layers by using more water in layers with adequate supplies. The potential water use must be adjusted for water deficits to obtain the actual use for each layer.

$$u_i = u_{pi}\,, \qquad SW_i > 0.25\,FC_i \tag{181}$$

$$u_i = u_{pi}\,(SW_i\,/\,0.25\,FC_i)\,, \qquad SW_i \leq 0.25\,FC_i \tag{182}$$

Note that the actual use for the layers above the i'th layer in Eq. 180 must be known. This presents no problems, however, since $u_j = 0$ when $i = 1$.

b. Temperature Stress.--The temperature stress factor is computed with the equation

$$TS = \exp\ \delta\,[\,(T_o - T)\,/\,T]^2 \tag{183}$$

where TS is the temperature stress factor (0-1), δ is the temperature stress parameter for the crop, T_o is the optimal temperature for the crop in °C (25° C for summer and 13° C for winter crops), and T is the daily average air temperature in °C. The stress parameter is evaluated by appropriate substituting and rearranging of Eq. 183.

$$\delta = \frac{\ln(0.9)}{\left(\dfrac{T_o - [\,(T_o + T_b)\,/\,2]}{(T_o + T_b)\,/\,2}\right)^2} \tag{184}$$

where T_b is the base temperature for the crop in °C (8° C for summer and 0° C for winter crops). Equation 184 sets $TS = 0.9$ when the air temperature is half way between T_b and T_o.

I. Tillage and Residue

The SWRRB tillage component was designed to partition the above-ground biomass at harvest. A portion of the biomass is incorporated into the soil while the remainder is left on the soil surface as residue. Once the residue is incorporated, it has no impact on the model. Also no change is made in bulk density due to tillage.

The residue is set at harvest according to one of the four tillage practices chosen.

Fall plow:

$$RSD = .05\ WLV \tag{185}$$

Spring plow:

$$RSD = 0.25\ WLV \tag{186}$$

Conservation tillage:

$$RSD = 0.50\ WLV \tag{187}$$

Zero tillage:

$$RSD = 0.95\ WLV \tag{188}$$

where RSD is the residue left after harvest in kg/ha and WLV is the above ground biomass minus crop yield at harvest in kg/ha.

The residue decays throughout the remainder of the year (harvest to next harvest) as a function of soil water content and soil temperature.

$$DECR = 0.05\ \min(CDG, SUT) \tag{189}$$

where DECR is the daily decrease in residue, and

$$RSD = RSD\ (1 - DECR) \tag{190}$$

where CDG and SUT are soil water and soil temperature residue decay factors.

$$CDG = \frac{0.9\ ST_2}{ST_2 + \exp(9.93 - 0.312\ ST_2)} \tag{191}$$

and

$$SUT = SW_2 / FC_2 \qquad (192)$$

where ST_2, SW_2, and FC_2 are the soil temperature, soil water content, and field capacity of the second soil layer.

J. Irrigation

The SWRRB user has the option to simulate dryland or irrigated agricultural areas. If irrigation is indicated, he must also specify the runoff ratio (volume of water leaving the field/volume applied) and a plant water stress level to start irrigation. The plant water stress factor ranges from 0 to 1.0 (1 means no stress and 0 means no growth). These stress factors are described in the Crop Growth section. When the user-specified stress level is reached, water is applied according to the equation

$$AIR = (FC - SW) / (1 - EFI) \qquad (193)$$

where FC is the root zone field capacity in mm, SW is the root zone water content before irrigation in mm, EFI is the runoff ratio, and AIR is the volume of irrigation water applied in mm.

Model Operation

User convenience was an important consideration in designing the SWRRB model. Some of the convenience features include the following: (1) Special input data forms were developed for use with an interactive data entry system. (2) Since SWRRB operates on a daily time step, computer run time is only about 50 seconds per year of simulation (for a basin with four subbasins) on a Compaq 386 computer with a 16 MHz microprocessor and an 80287 coprocessor. (3) The model can be run on a variety of computers since storage requirements are only 124K bytes.

Because of the provision for subdividin basins and because each subbasin can use a different rain gage, SWRRB is not limited by drainage area. The basins can be subdivided because of differences in soils, topography, land use, etc. The SWRRB code is dimensioned to handle a maximum of ten subbasins. Since all arrays are dimensioned in one file (SWRRB1.FOR), the array sizes can easily be increased allowing SWRRB to handle any number of subbasins. However, the interactive data entry system can only handle ten subbasins and the remaining data must be entered into the input file manually. Also, the increase in computer run time is almost directly proportional to the increase in subbasins. The vertical direction of the model is capable of working with any variation in soil properties--the soil profile is divided into a maximum of ten layers (the top layer thickness is set at 10 mm and all other layers may have variable thickness). A simplified SWRRB flowchart is shown on the following pages.

Some user options provided by SWRRB are:

1) Precipitation and temperature may be read in or generated independently. Solar radiation is always generated.

2) The same weather sequence may be repeated for any number of simulations at the same site or a new weather sequence may be generated for each simulation.

3) Daily, monthly, or yearly output may be specified.

4) The basin may be divided into a maximum of ten subbasins based on soils, land use, topography, vegetation, rainfall, temperature, etc.

5) A maximum of ten soil layers may be used.

6) The initial soil water state may be read in or estimated.

7) Ponds may be simulated.

8) Reservoirs may be simulated.

9) Dates of planting and harvesting are specified by user.

10) Any of four tillage operations may be specified for each subbasin (fall plow, spring plow, conservation tillage, and zero tillage).

11) Dryland agriculture or irrigation may be simulated.

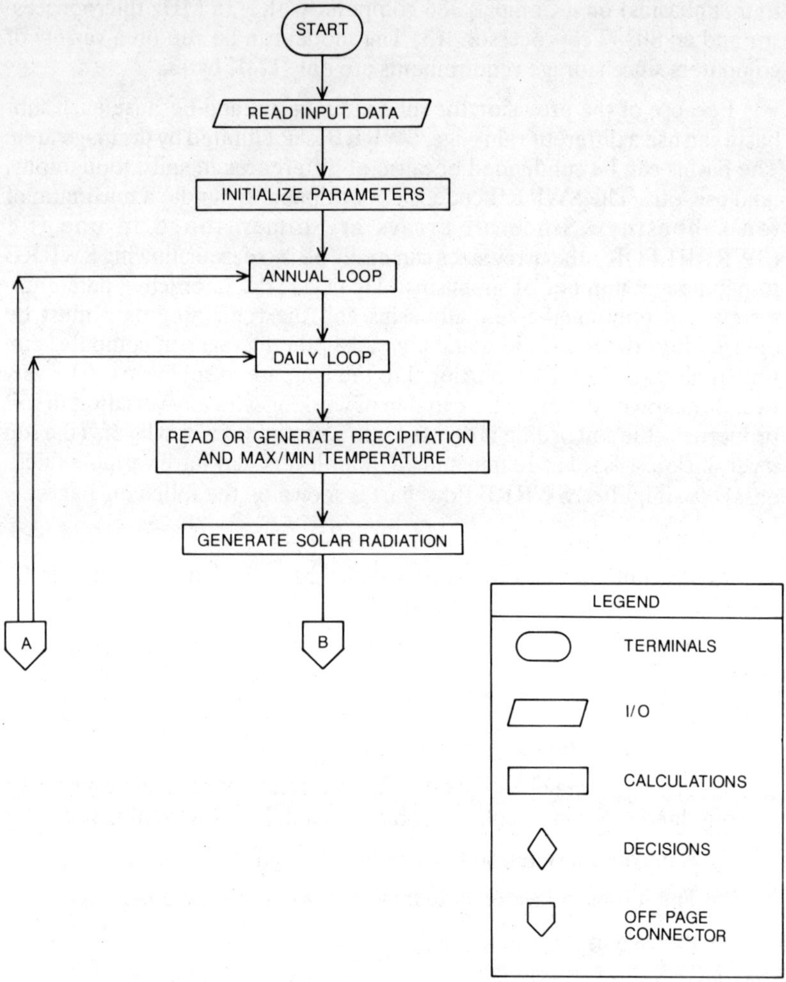

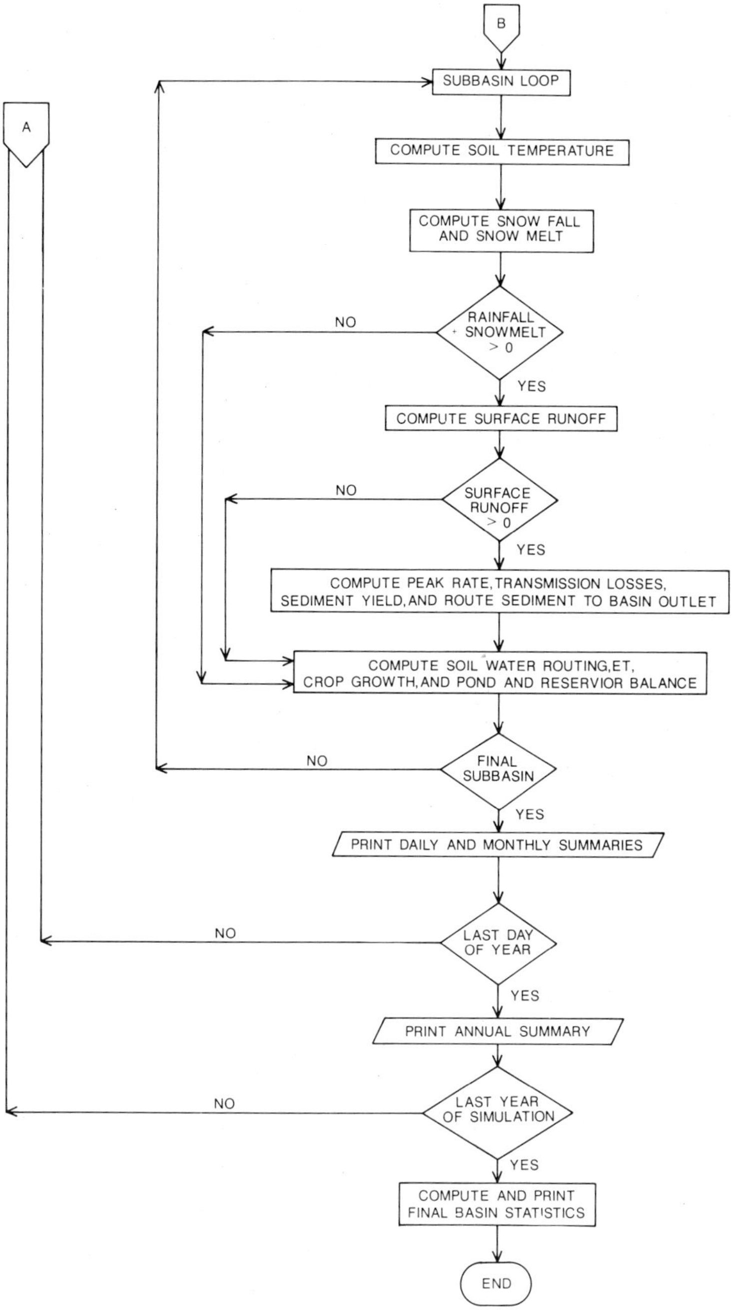
B
A
SUBBASIN LOOP
COMPUTE SOIL TEMPERATURE
COMPUTE SNOW FALL AND SNOW MELT
RAINFALL + SNOWMELT > 0
NO
YES
COMPUTE SURFACE RUNOFF
SURFACE RUNOFF > 0
NO
YES
COMPUTE PEAK RATE, TRANSMISSION LOSSES, SEDIMENT YIELD, AND ROUTE SEDIMENT TO BASIN OUTLET
COMPUTE SOIL WATER ROUTING, ET, CROP GROWTH, AND POND AND RESERVIOR BALANCE
FINAL SUBBASIN
NO
YES
PRINT DAILY AND MONTHLY SUMMARIES
LAST DAY OF YEAR
NO
YES
PRINT ANNUAL SUMMARY
LAST YEAR OF SIMULATION
NO
YES
COMPUTE AND PRINT FINAL BASIN STATISTICS
END

Subroutine Descriptions

The SWRRB main program and 36 subroutines contain approximately 2680 FORTRAN statements. The main program initializes variables, does all the input and output operations, controls the daily simulation of each subbasin by calling the appropriate subroutines, accumulates monthly and annual values, and computes average values for the simulation period. A brief description of the operation of each subroutine follows.

BLOCK DATA SUBROUTINE -- Initializes variables that are in common. It also contains tillage operation factors and generator seeds for the weather generator.

SUBROUTINE GCYCL3 -- Cycles the random number generator 0-100 times before simulation begins.

SUBROUTINE CLGEN4 -- Simulates daily solar radiation. Simulates daily precipitation and/or maximum and minimum air temperature at the user's option. Calls FUNCTIONS RANDN and DSTN1.

FUNCTION RANDN -- Provides random numbers ranging from 0 to 1.0.

FUNCTION DSTG -- Provides numbers from a gamma distribution, given two random numbers.

FUNCTION DSTN1 -- Computes the distance from the mean of a normal distribution with mean = 0 and standard deviation = 1, given two random numbers.

SUBROUTINE SNOM5 -- Predicts daily snow melt when the average air temperature exceeds 0°C.

SUBROUTINE CRPMD6 -- Predicts daily potential growth of total plant biomass and roots and calculates leaf area index. Incorporates residue for tillage functions and decays residue on ground surface. Calls SUBROUTINES SWU19 and TSTR7 and adjusts daily dry matter to water stress calculated in SWU19 and temperature stress from TSTR7.

SUBROUTINE TSTR7 -- Applies a temperature stress to crops.

SUBROUTINE EVAP8 -- Computes the amount of soil evaporation and the potential plant evaporation using Ritchie's ET model (Ritchie, 1972).

SUBROUTINE ALPH9 -- Computes α, a dimensionless parameter that expresses the fraction of total rainfall that occurs during the watershed time of concentration.

SUBROUTINE PKQ10 -- Computes the peak runoff rate for each subbasin and the entire basin using a modification of the Rational Formula (Williams et al., 1984).

SUBROUTINE VOLQ11 -- Predicts daily runoff given daily precipitation and snow melt using a modified SCS curve number approach.

SUBROUTINE RFLOW2 -- Calculates return flow travel times for each soil layer based on saturated conductivity and baseflow factors.

SUBROUTINE TRAN12 -- Computes transmission losses based on the procedure described in Chapter 19 of the SCS National Engineering Handbook (USDA, 1983).

SUBROUTINE YSED13 -- Predicts daily soil loss caused by water erosion using the Modified Universal Soil Loss Equation (Williams and Berndt, 1977).

SUBROUTINE RTSED1 -- Routes sediment from subbasin to basin outlets. Deposition is based on fall velocity and degradation on Bagnold's steam power concept.

SUBROUTINE POND14 -- Routes water and sediment through farm ponds. Also computes evaporation and seepage from the ponds.

SUBROUTINE RES15 -- Routes water and sediment through reservoirs. Computes evaporation and seepage from the reservoir.

SUBROUTINE PERC16 -- Computes percolation and lateral subsurface flow from a soil layer when field capacity is exceeded.

SUBROUTINE PKRN17 -- Computes percolation by crack flow.

SUBROUTINE PURK18 -- The master percolation component. It divides each layers flow into 4 mm slugs and manages the routing process. Calls PERC16 and PKRN17.

SUBROUTINE SWU19 -- Distributes potential plant evaporation through the root zone and calculates actual plant water use based on soil water availability. Also estimates water stress factor for use in CRPMD6.

SUBROUTINE SWBL20 -- Checks the soil water balance at the end of a simulation.

SUBROUTINE VBL21 -- Checks the water and sediment balance for ponds and reservoirs at the end of a simulation.

SUBROUTINE SOLT22 -- Estimates daily average temperature at the bottom of each soil layer.

SUBROUTINE XMON23 -- Determines the month, given the day of the year.

FUNCTION AHU -- Accumulates heat units to calculate the potential heat units for each crop before daily simulation begins.

FUNCTION JDT -- Computes the day of the year, given the month and day of the month.

FUNCTION ICL -- Computes the day of the month, given the julian day.

SUBROUTINE DELR24 -- Computes the within subbasin delivery ratios required to estimate sediment particle size distributions at the subbasin outlets.

SUBROUTINE VALD25 -- Controls the water and sediment yield validation by calling SUBROUTINE STAT27.

SUBROUTINE SORT26 -- Sorts arrays with high value first. Called by SUBROUTINE STAT27.

SUBROUTINE STAT27 -- Computes statistics on monthly measured and predicted water and sediment yields. Calls SUBROUTINE SORT26.

SUBROUTINE READ -- Reads general, weather, basin, routing, pond, and reservoir data.

SUBROUTINE OPEN -- Opens all input and output files for reading and writing.

Model Inputs

Two methods are available for inputting SWRRB data--batch and interactive. Appendix VIII contains a description of the SWRRB data requirements including variable definitions, formats, number of cards, units, typical range, and where to obtain the information. This table was designed to assist in assembling data for batch processing. Appendix IX contains data assembly forms designed for use with the interactive input system. The structure of the interactive system is similar to the one used for the EPIC (Erosion-Productivity Impact Calculator) model (Williams et al., 1986).

The following discussion expands the material presented in Appendix VIII to assist the new SWRRB user in assembling data. The data assembly forms and interactive data entry system should prove quite convenient for microcomputer users and mainframe users if a Time Sharing Option teleprocessing system is available. Since the input is the same for either system, an expanded description of the material in Appendix VIII will be helpful to the user regardless of choice of input systems. The numbers preceeding each item correspond with the numbers in column one of Appendix VIII.

(1) **TITLE.** The title cards can be used to describe location, soil types, land use, reasons for subdividing the basin, or any other information the user chooses. Three cards must be input for the title. If three cards are not needed, additional blank cards must be added.

(2) **PROGRAM CONTROL CODES.** The number of years of simulation (NBYR) can range from 1-100 years. However, 20-30 years may be adequate to estimate frequency distributions used to solve many problems. The beginning year (IYR) can be any number. The actual beginning year of record is used when simulating systems when measured data from the simulation period is used. Often it is convenient to begin hypothetical simulations on year one. The number of subdivisions can be made on the basis of soils, land use, vegetation, topography, temperature, rainfall, etc. There are no restrictions on where the subbasin lines may be drawn; however, each subbasin must include a channel reach. If necessary, the number of subbasins can be increased by increasing the size of the pertinent arrays in the computer code. The print code (IPD) allows the user to specify daily, monthly, or annual output. The annual printout minimizes the out-

put volume and is adequate for most long-term simulations (25-100 years). Monthly output provides necessary information for evaluating model performance within a growing season or for examining monthly runoff, erosion, etc. Thus, monthly output is normally used in short-term (1-20 year) simulations and is particularly useful in model testing. The main purpose of daily output is error detection, although it may be useful in simulating within growing season conditions. Caution should be exercised in specifying daily output for runs longer than one year to avoid excessive printout. The rainfall input code (NSIM) provides four user options. Most simulations used in decision making specify NSIM = 2 or NSIM = 4 (precipitation is generated for single and multiple rain gages, respectively) because the weather generator is reliable, gives realistic results, and is far more convenient than inputting precipitation. If precipitation is to be simulated for each subbasin (NSIM = 4), the X and Y-centroid coordinates (XIJ and YIJ) of each subbasin must be input. There are two options for inputting daily precipitaion data (NSIM = 1 for reading in a single gage, and NSIM = 3 for reading in one gage for each subbasin). These two options are mainly employed in model testing and other research studies. The temperature input code (MSIM) provides four user options and is independent of the precipitation input chosen. For most applications it will be sufficient to generate a single maximum and minimum temperature for the entire basin (MSIM = 2). Often in mountainous basins, there is a need to subdivide the valleys and the mountains because of differences in total precipitation, snowfall, vegetation, and temperature. Large errors in snowfall and snow melt can result if these temperature differences are not taken in account. In mountainous regions without available daily temperature data, it may be convenient to simulate temperature for each subbasin (MSIM = 4). Average monthly maximum and minimum temperatures are input for each subbasin. Realistic modeling of snowfall and snow melt can be made by generating temperature for each sub-basin. In all regions where snow is an important hydrologic factor and the simulation objective is to obtain good agreement between measured and predicted water and sediment yields, it may be necessary to input the daily temperature data (MSIM = 1 to read in a single maximum and minimum temperature for the entire basin and MSIM = 4 to read in maximum and minimum temperature for each subbasin). The random number generator code (IGN) indicates the number of times the random number generators are cycled before the simulation begins. The random number generator seeds (a set of random numbers) are contained in the SWRRB program data statements. If IGN = 0, the simulation begins with these seeds. Setting IGN0 allows the user to start each simulation with different seeds if desired--each time the generators cycle, they produce a new set of seeds. This feature is convenient for simulating several different weather sequences at a particular location. If measured monthly water and sediment yields are available, setting IWST = 1 (for water) and ISST = 1 (for sediment) allows to read in the measured values and compute relevant statistics on the measured and predicted values (i.e. means, standard deviations, R^2, regression slopes, etc.). If the measured values are not available, setting IWST and ISST to zero allows SWRRB to skip the statistical comparisons.

(3) **GENERAL DATA.** The first general data item is the watershed area (DA). Because of the provision for subdividing basins and because each subbasin can use different precipitation and temperature data, SWRRB is not limited by drainage area. When daily rainfall data is taken from a rain gage a considerable distance from the watershed, it may be necessary to use a rainfall correction factor (RF) other than one. If the difference between average annual precipitation at the watershed and rain gage is known, a rainfall correction factor from 0.5 to 1.5 can be used. RF is multiplied by the daily rainfall to obtain the corrected daily rainfall. The baseflow factor (BFF) is used in computing subsurface flow when return flow travel times are not input (Table III.4). Basin lag time (BRT) lags the subsurface flow. For BRT = 0 all subsurface flow reaches the subbasin outlets on the day it occurs. Judgment is required to set BRT to as many days as subsurface flow from a precipitation event is expected to contribute to streamflow. Initial soil water storage at the start of simulation is estimated as a fraction of field capacity (FFC). The field capacity soil water content (33 kPa) is multipled by FFC to obtain the initial soil water storage. If zero is entered FFC is calculated by SWRRB as a function of average annual precipitation.

(4) **X AND Y CENTROID COORDINATES.** The X and Y centroid coordinates (XIJ and YIJ) for each subbasin are input only if rainfall is simulated for multiple subbasins. Any reference point may be selected by the user.

(5) **GENERAL WEATHER DATA.** The TP-40 (Hershfield, 1961) 10-year frequency rainfall amount for 0.5 h (TP5) and 6 h (TP6) can be found in Table II.1 for 134 locations around the country. The number of years of maximum monthly 0.5 h rainfall available (TP24) can be obtained from the National Weather Service (U.S. Department of Commerce, 1979) and are also available for the 134 locations in Table II.1. The latitude of the water-

shed (YLT) is site specific, and thus user supplied. An example data set of weather generator parameters for Birmingham, AL, is shown in Table I.2.

(6) MONTHLY TEMPERATURE DATA. The weather generator forces long-term monthly maximum and minimum simulated air temperature to agree closely with the average monthly maximum (OBMX) and minimum (OBMN) air temperatures input to SWRRB. Values of OBMX and OBMN are also used to compute the mean amplitude of annual air temperature for use in the soil temperature model and crop growth model. Table I.1 contains OBMX and OBMN data. Data for other locations can be obtained from the Annual Summary of Climatological Data (U.S. Department of Commerce, 1978). If the temperature is simulated for each subbasin (MSIM = 4), values of OBMX and OBMN are input for each sub-basin . Temperatures are simulated for the first sub-basin and adjustments are made on the remaining sub-basins. This adjustment is the difference in OBMX or OBMN between the first and remaining sub-basins. Coefficient of variation for monthly temperature (CVT) is used with OBMX and OBMN in the weather generator to simulate daily maximum and minimum air temperatures. CVT can also be found in Table I.1.

(7) MONTHLY SOLAR RADIATION DATA. The weather generator also forces long-term monthly simulated solar radiation to agree closely with the average monthly solar radiation (OBSL). Table I.1 contains OBSL data. Monthly radiation data for other locations is available in Climates of the States (Water Information Center, Inc., 1974).

(8) MONTHLY RAINFALL DATA. The monthly maximum 0.5 h rainfall for the period of record (WI), is used in estimating the monthly maximum 0.5 h alpha. Alpha is a dimensionless parameter that expresses the fraction of total daily rainfall that occurs in 0.5 h, and is used in determining peak runoff rate. Table I.1 containes values of WI.

(9) PRECIPITATION GENERATION PARAMETER TABLE. The precipitation generator parameters (PRW_1 and PRW_2) are the monthly probabilities of a wet day after a dry day and of a wet day after a wet day, respectively. If PRW_1 and PRW_2 are unavailable, they can be estimated from the average number of days of precipitation in a month (WLV). The monthly mean precipitation for an event (RST_1), the monthly standard deviations of daily precipitation (RST_2), the monthly skew coefficients

(RST_3), PRW_1, and PRW_2 can be found in Table II.2. PRW_1, PRW_2, and RST_1 are also used to calculate the average days of precipitation for each month which is needed for estimating peak runoff rates.

(10) BASIN DATA. The fraction of the basin in each subbasin (FLU) can be obtained from topographic maps using a planimeter. If only one subbasin is needed, FLU = 1. The runoff curve number (CN2) is the SCS antecedant moisture condition 2 curve number (USDA, Soil Conservation Service, 1972). Table III.1 is the SCS table used in estimating CN2. Also, the hydrologic soil group is given in Appendix V for all soils in Soils-5. Soil albedo (SALB) values range fom 0.10 for dark colored soils, to 0.15 for light soils. Since there is little variation in soil albedo, SALB can be estimated from a general knowledge of the soil or a value of 0.125 can be assumed without significantly influencing evaporation. For long-term simulations used in decision making, the water content of snow on the ground at the beginning of simulation (SNO) is usually not known, but in most cases the estimate is not critical after the first year. If a measured value of SNO is available at the beginning of a simulation, it should be used. If SNO is not available, SWRRB will estimate it based on December temperature and precipitation. Effective hydraulic conductivity of the main channel alluvium (CHK1) and average main channel width (CHW1) are required for determining transmission losses (Table III.3). The channel length (CHL) is the distance along the channel from the sub-basin outlet to the most distant point in the subbasin. The average channel slope (CHS) is computed by dividing the difference in elevation between the subbasin outlet and the most distant point in the subbasin by CHL1. Both CHL and CHS can be obtained from topographic maps. Topographic maps can be purchased from the U.S. Geological Survey at the addresses shown in Appendix V. The channel "n" value (CHN) and overland flow "n" values are Manning's "n" values. Table III.2 contains suggested values of Manning's "n" for various conditions. Effective hydraulic conductivity of the channel alluvium (CHK) is given in Table III.3 for various channel bed material. Transmission losses are based on CHK. Return flow travel time (RT) is the time required for subsurface flow from the centroid of the subbasin to reach the subbasin outlet. The value of RT is input for each subbasin by the SWRRB user or can be calculated by SWRRB from soil hydraulic properties and flow characteristics. Experienced hydrologists familiar with the base flow characteristics of watersheds within a region should have little problem in assigning reasonable values to RT. However, if the user is not familiar with the watershed, SWRRB will estimate RT based on the soil's saturated conductivity and a parameter called the baseflow factor (BFF). Boughton (1984) developed BFF for a simple water yield model. Values of BFF are found in Table III.4. Sediment concentration in return flow (CSS) is usually very low and does not contribute significantly to total sediment yields unless return flow is very high. Measurements of CSS at Riesel, Texas, ranged from 300-700 ppm. Unless the user is aware of unusual situations where CSS may be extremely high, a value of 500 ppm is a good estimate and will yield realistic results. Values of the USLE erosion control practice factor (EP) provided by Wischmeier and Smith (1978) are contained in Table IV.1.

The average land slope of a subbasin can be estimated with the Grid-Contour Method (Williams and Berndt, 1976) using the equations

$$S_d = (N_d)(CI) / LG_d \tag{194}$$

$$S = \sqrt{S_L^2 + S_T^2} \tag{195}$$

where S_d is the slope in one grid direction in m/m, N_d is the total number of contour crossings for grid lines in direction d, CI is the contour interval, LG_d is total length of grid lines within the subbasin in direction d in mm, S is the average slope of the subbasin in m/m, S_L is the slope in the length grid direction obtained from Eq. 194, and S_T is the slope in the width direction in m/m.

The average slope length is estimated for each subbasin with the Contour-Extreme Point Method (Williams and Berndt, 1976)

$$\lambda = LC / 2\,EP \tag{196}$$

where EP is the number of extreme points (channel crossings) on the contours, LC is the total length of all contours within the subbasin in m, andΠ10M λ is the average slope length in m/m.

(11) ROUTING DATA. If the basin is not subdivided, routing data is not required. Average channel depth (CHD), average channel width (CHW2), channel slope (CHSS), channel length (CHL2), and channel "n" (CHNN), from the subbasin outlet to the basin outlet, are used to calculate travel times using Manning's equation. The routing data is used in sediment routing and computing transmission losses. CHSS and CHL2 can be obtained from topographic maps (see Appendix VI). If detailed channel cross-section data is unavailable, CHD and CHW may be estimated from a general knowledge of the basin. Table III.2 contains CHNN. Effective hydraulic conductivity of the channel alluvium from the subbasin outlet to the basin outlet (CHK2) is needed for determining transmission losses (Table III.3). Also, USLE soil erodibility and crop management factors (CHXK and CHC) for the channel are needed for sediment routing.

(12) POND DATA. The fraction of each subbasin that flows into ponds (FP) can be determined from topographic maps. The total surface area of all ponds (SAX) and runoff volume from the pond catchment area to fill the ponds (VMX) can be obtained from local SCS offices. The SCS has general design volumes for various regions (i.e. one inch in central Oklahoma). Initial pond volumes (VM) at the start of simulation range from 0 to VMX. If VM is unknown, a value of 0.75*VMX is adequate since all simulations start on January 1. Initial sediment concentration (CS) and normal sediment concentration (CFP) in ponds are usually very low (200-700 ppm). Unless more detailed information is known, a value of 400 ppm is a reasonable estimate. The hydraulic conductivity of the pond bottoms (HC) controls the amount of seepage from the ponds and is generally very low. A value of 0.08 mm/h has worked well in preliminary studies. Values of HC larger than 0.08 mm/h cause excessive seepage for most situations.

(13) RESERVOIR DATA. Reservoir design data is needed to complete the required SWRRB reservoir data. It is usually necessary to consult the designer or operator of the reservoir (i.e. Corps of Engineers, Soil Conservation Service, etc.) to obtain the necessary inputs. As with ponds, initial sediment concentration (CSR) and normal sediment concentration (CFR) in reservoirs is very low (400 ppm). A value of 0.08 mm/h for the hydraulic conductivity of reservoir bottoms (HCR) also proved adequate.

(14) SOILS DATA. The number of different soil series (NSO) is input by the user. More than one subbasin could use the same soil series or each subbasin could have a different series. For example, there could be four subbasins and only two different soil series. Next, the soil series that corresponds to each subbasin (MS) is input. All remaining soils data for SWRRB can be taken from the SCS Soils-5 data base. The database is compiled from the SCS Soils-5 Interpretation Records, prepared by SCS staff, which provide information on the characteristics and interpretive properties of all soils series identified in the United States. The data has been systematically collected by SCS for decades and contains the properties and characteristics of more than 14,000 soils. It is maintained and updated by the SCS at the Statistical Laboratory of Iowa State University in Ames, Iowa. The database contains a comprehensive listing of soil properties and land use capabilities. The specific soils data required for input to SWRRB was extracted to form a much smaller database. An example data set for the Ripple soil series is shown in Table V.1. Table V.2 gives various sources to obtain the necessary Soils-5 data for SWRRB. Probably the most convenient access to Soils-5 is direct access from the interactive data entry program (Table V.4).

There is often more than one surface layer given. In most cases the first layer shown is the predominant layer. The Soils-5 sources give ranges for most of the soil parameters. Unless more detailed information about the soil is known, it is usually best to use the average of the range. All soil layer depths may be assigned by the user except the top layer (it is always 10 mm thick). Subsurface flow is sensitive to soil layer thickness. We recommend a maximum thickness of 500 mm to maintain reliable subsurface flow predictions. Subsurface flow could be underestimated if the recommended thickness is exceeded. Also, the subsurface flow model is extremely sensitive to lower layers that are impermeable. If an impermeable layer is suspected (such as impermeable bedrock) that is not described in the soil profile description, it should be added to the soil input data. The number of soil layers (NS1) is also assigned by the user. Usually the depths to the bottom of the layers (Z) are assigned to coincide with the Soils-5 data. Up to 10 layers are allowed. Bulk density (BD) and available water capacity (AWC) are both given as ranges in Soils-5. If these values are missing in Soils-5, mean values from Table V.3 based on texture may be used (Svetlosanov and Knisel, 1981). Total porosity (POR) is calculated using bulk density

$$POR = 1 - (BD / 2.65) \tag{197}$$

The saturated conductivity (SC) is taken from the Soils-5 permeability. Clay content (CLA) and percent passing #200 sieve (SIL) are used to determine sand, silt, and clay contents. Silt content is calculated as the difference between SIL and CLA. Sand content (SAN) is then simply SAN = 100-SIL-CLA. Wilting point moisture content (WP) is computed with the equation

$$WP = 0.4\,(BD)\,(CLA) \tag{198}$$

Available water capacity (AWC) is added to the wilting point moisture content to obtain the field capacity. The USLE soils K factor (EK) can also be found in the Soils-5 data base. The detached sediment particle size distribution can be input by the user or estimated by SWRRB. If the distribution is input, the fraction of sand (.20 mm), silt (.01 mm), clay (.002 mm), small aggregates (.03 mm), and large aggregates (.50 mm) are required.

(15) CROP DATA. Planting (IPL) and harvest (IHV) dates are input to SWRRB. Annual (IRD = 1) or perennial (IRD = 2) vegetation must be specified. Root growth is allowed to begin at planting for annual crops while perennial crops maintain their roots throughout the growing season. There are four tillage operations: ITIL = 1 for fall plow, ITIL = 2 for spring plow, ITIL = 3 for conservation tillage, and ITIL = 4 for zero tillage. The amount of residue left after harvest (above ground biomass minus yield) for each operation is 0.05, 0.25, 0.50, and 1.0, respectively. For row crops such as corn, soybeans, grain sorghum, sunflowers, cotton, and peanuts, the average annual C factor (CVA) for water erosion is 0.32. A value of 0.12 can be used for drilled crops such as wheat, oats, or barley. The CVA values for alfalfa and pasture are 0.04 and 0.008, respectively. Forested areas generally have CVA values that range between 0.04 and 0.008, unless grazing or other management practices would cause the CVA value to be somewhat higher. Values of CVA for other conditions can be found in Wischmeier and Smith (1978). Maximum leaf area index (BLAI) is assumed to equal 5 for all vegetation when average annual rainfall is sufficient. This includes all areas east of the Mississippi River and any areas that are irrigated. Most western rangeland has BLAI values less than 5. Figure VII.1 gives typical BLAI values for six major land resource regions (MLRR) which contain most of the rangeland in the U.S. These estimates were obtained by comparing biomass production for a number of soils in each MLRR. These values are rough estimates for the entire MLRR and are not site specific. In SWRRB, BLAI affects evapotranspiration and daily biomass production. RZ is the maximum plant rooting depth. By default (entering zero for RZ), SWRRB assumes RZ to equal the depth of the bottom soil layer.

(16) IRRIGATION DATA. The irrigation code (IRR) is set to one if irrigation is applied to a subbasin. The value of the plant water stress factor when irrigation begins (WSF) provides flexibility in irrigation scheduling and in the total volume of water applied. For example, if only supplemental irrigation is desired, WSF should be set to a low value ($\approx$0.1). The low value allows the plants to be stressed severely before water is applied. Conversely, WSF should be set to a high level ($\approx$0.9) to assure little plant water stress (typical of drip irrigation). Most irrigation systems can be simulated satisfactorily using a WSF value of 0.5. The irrigation runoff ratio (EFI) is the ratio of the amount of runoff to the total amount of water applied.

(17) DAILY RAINFALL AND TEMPERATURE DATA. No input is required if rainfall and temperature are simulated. If daily rainfall or temperature is read in, one card per day for the number of days of simulation is used. If rainfall in input, list daily rainfall (mm) for each subbasin (NSIM = 3), or one raingage for the entire basin (NSIM = 1). Daily temperature (°C) is contained in a separate file. If MSIM = 3, list one maximum and minimum temperature on each card. List maximum temperature for each subbasin, and then minimum temperature for each subbasin on the same card for MSIM = 3. The format is FORMAT(5X,15F5.1).

Interactive Data Entry System

The Interactive Data Entry System consists of the Enteract program, Prompts file, explanation file, weather parameter file, and soils files. Enteract is designed to build a SWRRB data file by accepting input from the user, editing, and then writing it into a new file in the correct SWRRB format. The basic operation of the program uses this logic to remove as many input errors as possible. It builds each line of the data file one variable at a time. The values are accepted and checked against the ranges that are read in from the Prompts file. If the values are within the acceptable limits, Enteract continues to the next prompt. If the line has been completed, a prompt appears on the screen to ask if the last value is correct and ready to write in the data file. This is done to give the user a chance to correct the last value before it is written into the new data file. After a line is written, the values cannot be changed by Enteract. However, at any time before the line is written, the user may enter a value of "-999" and Enteract will back up one prompt. This is often useful when an incorrect value that is within the acceptable range has been entered and written. Of course, the user can always edit the new data file after Enteract is finished.

In some cases, such as monthly radiation data, 12 values are accepted on one line in "free format." In "free format" all the values may be entered on one line if they are separated by at least one space or comma. This speeds up the entry of values having the same range and description; however, if one of the 12 values is out of range, all 12 values will have to be re-entered. If carriage return is hit after only 11 of the 12 values have been entered, Enteract will not write to the new data file until all of the 12 values are entered. At this point, only the final value needs to be entered followed by carriage return. Entering an illegal value (such as 23.5.2 or an alphabetic character) will require the user to re-enter the values.

In the General Weather Data section (section 5), the user will be prompted to input the weather parameters or have the parameters entered automatically. If automatic entry is chosen, the user will refer to the weather generator parameters in Appendix II and enter the number corresponding to the weather station chosen. Enteract will format and write it to the SWRRB input file. If automatic entry is not chosen, the user inputs the values as prompted.

There is a similar option to automatically enter the soils data. In this case, the user will have the option to simply enter the soil series number in Appendix V (Table V.4) and Enteract will automatically format and write the corresponding Soils-5 data to the SWRRB input file. The user always has the option to override the soils data if necessary.

Running the Model on a Personal Computer

SWRRB and the Enteract program run on most IBM-PC compatibles and require less than 256K RAM memory. The diskettes can be read from any 360K byte drive. The programs are written in FORTRAN and can be compiled with any ANSI 77 FORTRAN compiler.

To run Enteract, simply type in "RUNENTR." The input data is written to a file called NEWDATA. NEWDATA must be renamed before another data set is assembled by Enteract, or the old file will be written over.

Once a data set has been assembled, SWRRB is ready to run. The file CONTROL.DAT controls the input/output of SWRRB. The first line contains the name of the SWRRB data file to be run. The second line contains the name of the measured maximum and minimum temperature file. The file containing the measured monthly water and sediment yield is found on the third line. The fourth line can be the printer's device name (i.e. LPT1), a file name to be printed at a later time, or viewed from the console (CON).

EXAMPLE CONTROL.DAT FILE

Line			
1	INFIL1	:	SWOKW522.DAT
2	INFIL2	:	SWOKW522.TEM
3	INFIL3	:	W522QY.DAT
4	OUTFIL	:	CON

File Descriptions

Files for SWRRB (PC Version)

SWRRB1.FOR -- SWRRB source file (common blocks)
SWRRB2.FOR -- SWRRB source file (block data)
SWRRB3.FOR -- SWRRB source file (main program)
SWRRB4.FOR -- SWRRB source file (subroutines)
SWRRB5.FOR -- SWRRB source file (subroutines)
SWOKW522.DAT -- Sample SWRRB dataset (Chickasha, OK)
SWOKW522.TEM -- Temperature data file (optional) (Chickasha, OK)
W522QY.STA -- Monthly measured water and sediment data file (optional)
CONTROL.DAT -- File controlling SWRRB input/output
RUNSWRRB.EXE -- The execute file to run SWRRB

Files for Enteract (PC Version)

ENTRSWRB.FOR -- Enteract source (main program and subroutines)
ENTRFILE.DAT -- File controlling Enteract input/output
WEATHER.DAT -- File containing the weather generation parameters
SOIL01.DAT -- File containing the Soils-5 data (A - CL)
SOIL02.DAT -- File containing the Soils-5 data (CO - G)
SOIL03.DAT -- File containing the Soils-5 data (H-MAR)
SOIL04.DAT -- File containing the Soils-5 data (MAS - RA)
SOIL05.DAT -- File containing the Soils-5 data (RE - T)
SOIL06.DAT -- File containing the Soils-5 data (U - Z)
PROMPTS.DAT -- File read by Enteract containing ranges for variables
NEWDATA -- Complete SWRRB dataset Enteract writes to
EXPLAIN.TXT -- File read by Enteract containing explanations for selected variables
RUNENTR.EXE -- The execute file to run Enteract

Example Applications

This section contains example input forms for use with the interactive data entry program. Three widely-varying examples were selected in an attempt to demonstrate some of the input options. If the basin is subdivided, a map of the basin showing the divisions is included. The divisions may be based on soils, land use, tillage operations, elevation, temperature, rainfall, etc. Following each set of input forms is a listing of the data prepared by the interactive entry program. The input data could have been prepared directly following instructions in Appendix VIII.

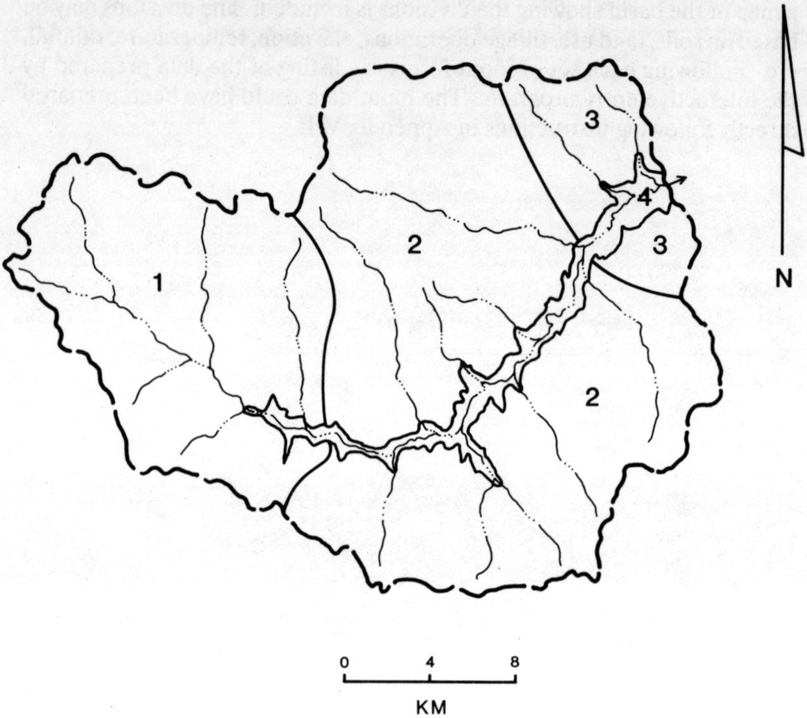

Watershed 522 at Chickasha, Oklahoma

*** SWRRB DATA ASSEMBLY FORM 1 ***

1. **TITLE:** Chickasha, Oklahoma
Watershed 522
Little Washita River Basin

2. **PROGRAM CONTROL CODES**

Number of years of runoff simulation 3

Beginning year of runoff simulation 1970

Number of subareas in basin 4

Printout frequency

Monthly (0) X Daily (1) ____ Annual (2) ____

Will the rainfall be:

Read in single rain gage for entire basin	(1)	
Simulated single rain gage for entire basin	(2)	
Read in one rain gage for each sub-basin	(3)	X
Simulated for multiple rain gages	(4)	

Will the max & min temperatures be:

Read in single max & min for entire basin	(1)	
Simulated single max & min for entire basin	(2)	X
Read in max & min for each sub-basin	(3)	
Simulated for each sub-basin	(4)	

Number of times random number generator cycles before simulation begins 0

Code for stat collection on monthly water yield

Skip collection	(0)	
Collect	(1)	X

Code for stat collection on monthly sediment yield

Skip collection	(0)	
Collect	(1)	X

*** SWRRB DATA ASSEMBLY FORM 2 ***

3. GENERAL DATA

Basin area (km^2) 538.2

Rainfall correction factor 1.

Baseflow factor 0.5

Basin lag time (days) 30.

Initial soil water storage -- fraction of field capacity 0

(entering 0. allows SWRRB to estimate it)

4. CENTROID COORDINATES OF SUB-BASINS

X: 9.34 12.29 22.11 12.29 ____ Y: 28.18 19.90 24.21 11.03 ____

5. GENERAL WEATHER DATA

If automatic entry is used, enter:

Weather station No. (Appendix I) ________ and skip to Section 10. If not, continue

TP-40 10-year frequency 0.5 hour rainfall (mm) 56.0

TP-40 10-year frequency 6.0 hour rainfall (mm) 108.0

Number of years for maximum monthly 0.5 hour rainfall record 8.0

Latitude of watershed (deg) 35.1

6A. MONTHLY TEMPERATURE DATA

Average monthly maximum air temperature (°C)

8.94 11.6 16.9 23.4 27.0 31.6 34.3 32.8 28.4 23.6 16.5 10.8

Average monthly minimum air temperature (°C)

-3.48 -1.78 3.23 9.77 14.01 19.3 21.7 20.3 16.9 9.82 3.58 -1.46

Coefficient of variation for monthly temperature

.34 .24 .22 .15 .12 .07 .06 .06 .11 .14 .19 .27

*** SWRRB DATA ASSEMBLY FORM 3 ***

6B. MONTHLY TEMPERATURE DATA FOR REMAINING SUB-BASINS

(SKIP UNLESS TEMPERATURE IS SIMULATED FOR EACH SUB-BASIN)

2 Maximum (°C)

3

4

5

2 Minimum (°C)

3

4

5

7. MONTHLY SOLAR RADIATION DATA

Monthly average daily solar radiation (ly)

220	291	370	453	511	553	554	501	392	332	237	197

8. MONTHLY RAINFALL DATA

Monthly maximum 0.5 h rainfall (mm)

8.1	11.4	28.4	25.1	42.4	55.4	39.1	28.4	31.7	22.9	26.7	13.5

9. PRECIPITATION GENERATION PARAMETER TABLE

(Enter P(W/D) & P(W/W) or DAYPR)

P(W/D)

.08	.12	.13	.16	.18	.22	.12	.18	.20	.12	.12	.13

P(W/W)

.30	.33	.30	.42	.43	.38	.35	.36	.33	.39	.34	.34

Days Prec

0	0	0	0	0	0	0	0	0	0	0	0

Mean

5.56	5.49	8.36	9.17	14.35	11.28	10.69	9.91	16.50	9.52	8.79	4.49

St. dev.

7.72	6.81	10.08	11.38	21.26	15.70	13.66	15.01	27.48	10.97	13.82	6.96

Skew coefficient

1.13	0.86	1.22	0.72	2.34	2.95	1.49	2.81	3.86	0.38	2.72	2.63

*** SWRRB DATA ASSEMBLY FORM 4 ***

10. BASIN DATA

Sub-basin:	1	2	3	4	5	Entire b
Fraction of basin in each subarea	.297	.564	.085	.054		
SCS runoff curve number - Condition II	77	75	77	77		
Soil albedo	.15	.15	.15	.15		
Water content of snow on ground at start of simulation (mm)	0	0	0	0		
Main channel length in each sub-basin (km) of simulation (mm)	14.2	11.7	6.8	27.1		41.2
Average channel slope in each sub-basin (m/m)	.0038	.0038	.0076	.0114		.001
Average main channel width (m)	9	11	11	13		
Effective hydraulic conductivity of channel alluvium (mm/hr)	10.	10.	10.	10.		
Channel N value	.05	.05	.05	.05		.05
Overland flow N value	.05	.05	.05	.05		.05
Return flow travel time (days) (skip if unknown)	0	0	0	0		
Sediment concentration in return flow (ppm)	750	750	750	750		
USLE erosion control practice factor P	1.0	1.0	1.0	1.0		
Average slope length for each sub-basin (m)	55	55	50	100		60
Average slope steepness for each sub-basin (m/m)	.08	.09	.08	.01		.08

11. ROUTING DATA (If only one sub-basin is used, skip to Section 12)

(All routing data is from sub-basin to basin outlet)

	1	2	3	4	5
Average channel width (m)	9.0	9.0	11.0	0	
Average channel depth (m)	2.0	2.0	2.5	0	
Channel slope (m/m)	.011	.011	.011	0	
Channel length (km)	27.1	12.3	2.1	0	
Channel N value (mm/hr)	.05	.05	.05	0	
Effective hydraulic conductivity of channel alluvium (mm/hr)	10.0	10.0	10.0	0	
USLE soils K factor for channel	.305	.305	.305	0	
USLE C factor for channel	1.00	1.00	1.00	0	

*** SWRRB DATA ASSEMBLY FORM 5 ***

12. POND DATA

(If not used skip to Section 13)

Sub-basin:	1	2	3	4	5
Fraction of each sub-basin that flows into ponds	.195	.195	.195	.195	
Total surface area of all ponds in each sub-basin (ha)	14	267	102	25	
Runoff volume from pond catchment area to fill ponds (mm)	24	24	24	24	
Initial pond volumes (mm)	20	20	20	20	
Initial sediment concentration in ponds (ppm)	400	300	450	250	
Normal sediment concentration in ponds (ppm)	400	300	450	250	
Hydraulic conductivity of pond bottoms (mm/hr)	.08	.08	.08	.08	

13. RESERVOIR DATA (If not used skip to Section 14)

	1	2	3	4	5
Fraction of each sub-basin that flows in reservoirs	.142	.293	.419	0	
Total reservoir surface area at emergency spillway (ha)	258	449	144	0	
Runoff volume from reservoir catchment area required to fill emergency spillway (mm)	89	89	89	0	
Total reservoir surface area at principal spillway (ha)	74	144	28	0	
Runoff required to fill to principal spillway (mm)	29.1	23.8	27.3	0	
Initial reservoir volumes (mm)	24	18	22	0	
Average principal spillway release rate ($m^3/s/km^2$)	.137	.117	.121	0	
Initial sediment concentration in reservoirs (ppm)	200	200	200	0	
Normal sediment concentration in reservoirs (ppm)	250	250	250	0	
Hydraulic conductivity of reservoir bottoms (mm/hr)	.08	.08	.08	0	

*** SWRRB DATA ASSEMBLY FORM 6 ***

14. SOILS DATA

Number of soils 4

Soil Series name

#1 Vanoss #2 Daugherty #3 Renfrow #4 Port #5 ______

Subbasin:	1	2	3	4	5
Soil Series for each subbasin:	1	2	3	4	
No. layers for each soil series (if automatic entry is used, skip to Section 15)	7	7	7	7	

Soil Series name Vanoss K Factor .31 Pass #200 Sieve 81

Layer:

	1	2	3	4	5	6	7	8	9	10
Depth to bottom layer (mm)	10	150	510	840	1190	1520	1900			
Bulk density (m/m^3)	1.40	1.40	1.55	1.58	1.55	1.55	1.55			
Available water capacity (mm/mm)	.19	.19	.19	.19	.18	.18	.18			
Saturated conductivity (mm/h)	16	16	15	12	12	12	12			
Clay content (%)	16	16	23	35	35	30	30			

Sediment size distribution (fraction-if unknown, skip to next soil)

Sand (.20 mm)	Silt (.01 mm)	Clay (.002 mm)	Sm Ag (.03 mm)	L Ag (.50 mm)

Maximum rooting depth (mm) ______

*** SWRRB DATA ASSEMBLY FORM 6 ***

14. SOILS DATA

Number of soils __________

Soil Series name

#1 __________ #2 __________ #3 __________ #4 __________ #5 __________

Subbasin: 1 2 3 4 5

Soil Series for each subbasin:

__________ __________ __________ __________ __________

No. layers for each soil series (if automatic entry is used, skip to Section 15)

__________ __________ __________ __________ __________

Soil Series name Daugherty K Factor .25 Pass #200 Sieve 50

Layer:

	1	2	3	4	5	6	7	8	9	10
Depth to bottom layer (mm)	10	150	560	760	970	1420	1830			
Bulk density (m/m^3)	1.42	1.42	1.42	1.60	1.60	1.60	1.60			
Available water capacity (mm/mm)	.13	.13	.08	.11	.13	.09	.08			
Saturated conductivity (mm/h)	50	50	50	21	21	21	21			
Clay content (%)	15	19	19	19	19	9	9			

Sediment size distribution (fraction-if unknown, skip to next soil)

Sand (.20 mm)	Silt (.01 mm)	Clay (.002 mm)	Sm Ag (.03 mm)	L Ag (.50 mm)

Maximum rooting depth (mm) __________

*** SWRRB DATA ASSEMBLY FORM 6 ***

14. SOILS DATA

Number of soils ___________

Soil Series name

#1 ___________ #2 ___________ #3 ___________ #4 ___________ #5 ___________

Subbasin:	1	2	3	4	5
Soil Series for each subbasin:					
No. layers for each soil series (if automatic entry is used, skip to Section 15)					

Soil Series name Renfrow K Factor .34 Pass #200 Sieve 81

Layer:

	1	2	3	4	5	6	7	8	9	10
Depth to bottom layer (mm)	10	280	330	530	760	1140	1905			
Bulk density (m/m^3)	1.42	1.42	1.57	1.55	1.55	1.55	1.55			
Available water capacity (mm/mm)	.15	.15	.15	.14	.14	.14	.14			
Saturated conductivity (mm/h)	7	5	.75	.75	.75	.75	.75			
Clay content (%)	23	23	34	45	44	42	42			

Sediment size distribution (fraction-if unknown, skip to next soil)

Sand (.20 mm)	Silt (.01 mm)	Clay (.002 mm)	Sm Ag (.03 mm)	L Ag (.50 mm)

Maximum rooting depth (mm) ___________

*** SWRRB DATA ASSEMBLY FORM 6 ***

14. SOILS DATA

Number of soils ___________

Soil Series name

#1 ___________ #2 ___________ #3 ___________ #4 ___________ #5 ___________

Subbasin: 1 2 3 4 5

Soil Series for each subbasin:

___________ ___________ ___________ ___________ ___________

No. layers for each soil series (if automatic entry is used, skip to Section 15)

___________ ___________ ___________ ___________ ___________

Soil Series name Port K Factor .30 Pass #200 Sieve 84

Layer:

	1	2	3	4	5	6	7	8	9	10
Depth to bottom layer (mm)	10	280	560	860	1270	1500	1828			
Bulk density (m/m^3)	1.42	1.42	1.42	1.45	1.45	1.45	1.45			
Available water capacity (mm/mm)	.19	.18	.19	.18	.18	.18	.18			
Saturated conductivity (mm/h)	12	12	12	12	12	12	12			
Clay content (%)	12	23	39	44	42	42	42			

Sediment size distribution (fraction-if unknown, skip to next soil)

Sand (.20 mm)	Silt (.01 mm)	Clay (.002 mm)	Sm Ag (.03 mm)	L Ag (.50 mm)

Maximum rooting depth (mm) ___________

*** SWRRB DATA ASSEMBLY FORM 7 ***

Subbasin 1

15. CROP DATA

Planting date: Month 3 Day 15

Harvest date: Month 9 Day 15

(Use start and end of growing season for perennial vegetation)

Vegetation: Annual (1) X Perennial (2) ____

Tillage Operation:

- Fall plow (1) X
- Spring plow (2) ____
- Conservation tillage (3) ____
- Zero tillage (4) ____

Average annual C-factor .17

Maximum leaf area index 4.0

Initial residue cover (kg/ha) ____

16. IRRIGATION DATA

Are the crops irrigated Yes (1) ____ No (0) X

Water stress when irrigation is applied ____

Irrigation runoff ratio ____

17. DAILY RAINFALL AND TEMPERATURE DATA

If rainfall is input, list daily rainfall (mm) for each subbasin.

If temperature is input, list maximum and minimum temperature for each subbasin in a separate file.

[Format (5X, 15F5.1)]

*** SWRRB DATA ASSEMBLY FORM 7 ***

Subbasin ___2___

15. CROP DATA

Planting date: Month ___3___ Day ___15___

Harvest date: Month ___9___ Day ___15___

(Use start and end of growing season for perennial vegetation)

Vegetation: Annual (1) ___X___ Perennial (2) ______

Tillage Operation:

- Fall plow (1) ______
- Spring plow (2) ___X___
- Conservation tillage (3) ______
- Zero tillage (4) ______

Average annual C-factor ___.13___

Maximum leaf area index ___4.0___

Initial residue cover (kg/ha) ______

16. IRRIGATION DATA

Are the crops irrigated Yes (1) ______ No (0) ___X___

Water stress when irrigation is applied ______

Irrigation runoff ratio ______

17. DAILY RAINFALL AND TEMPERATURE DATA

If rainfall is input, list daily rainfall (mm) for each subbasin.

If temperature is input, list maximum and minimum temperature for each subbasin in a separate file.

[Format (5X, 15F5.1)]

*** SWRRB DATA ASSEMBLY FORM 7 ***

Subbasin ___3___

15. CROP DATA

Planting date: Month ___3___ Day ___15___

Harvest date: Month ___9___ Day ___15___

(Use start and end of growing season for perennial vegetation)

Vegetation: Annual (1) ___X___ Perennial (2) ______

Tillage Operation:

- Fall plow (1) ______
- Spring plow (2) ______
- Conservation tillage (3) ___X___
- Zero tillage (4) ______

Average annual C-factor ___.10___

Maximum leaf area index ___4.0___

Initial residue cover (kg/ha) ______

16. IRRIGATION DATA

Are the crops irrigated Yes (1) ______ No (0) ___X___

Water stress when irrigation is applied ______

Irrigation runoff ratio ______

17. DAILY RAINFALL AND TEMPERATURE DATA

If rainfall is input, list daily rainfall (mm) for each subbasin.

If temperature is input, list maximum and minimum temperature for each subbasin in a separate file.

[Format (5X, 15F5.1)]

*** SWRRB DATA ASSEMBLY FORM 7 ***

Subbasin ___4___

15. CROP DATA

Planting date: Month ___3___ Day ___15___

Harvest date: Month ___9___ Day ___15___

(Use start and end of growing season for perennial vegetation)

Vegetation: Annual (1) ___X___ Perennial (2) ______

Tillage Operation:

- Fall plow (1) ______
- Spring plow (2) ______
- Conservation tillage (3) ______
- Zero tillage (4) ___X___

Average annual C-factor ___.05___

Maximum leaf area index ___4.0___

Initial residue cover (kg/ha) ______

16. IRRIGATION DATA

Are the crops irrigated Yes (1) ______ No (0) ___X___

Water stress when irrigation is applied ______

Irrigation runoff ratio ______

17. DAILY RAINFALL AND TEMPERATURE DATA

If rainfall is input, list daily rainfall (mm) for each subbasin.

If temperature is input, list maximum and minimum temperature for each subbasin in a separate file.

[Format (5X, 15F5.1)]

```
 1.   CHICKASHA, OKLAHOMA
 2.   WATERSHED 522
 3.   LITTLE WASHITA RIVER BASIN
 4.     31970   4   0   3   2   0   1   1
 5.   538.200   1.000    .500  30.000    .000
 6.      9.34   12.29   22.11   12.29   28.18   19.90   24.21   11.03
 7.     56.00  108.00    8.00   35.10
 8.    8.94 11.60 16.90 23.40 27.00 31.60 34.30 32.80 28.40 23.60 16.50 10.80
 9.   -3.48 -1.78  3.23  9.77 14.01 19.30 21.70 20.30 16.90  9.82  3.58 -1.46
10.     .34   .24   .22   .15   .12   .07   .06   .06   .11   .14   .19   .27
11.    220.  291.  370.  453.  511.  554.  544.  501.  392.  332.  237.  197.
12.    8.10 11.40 28.40 25.10 42.40 55.40 39.10 28.40 31.70 22.90 26.70 13.50
13.     .08   .12   .13   .16   .18   .22   .12   .18   .20   .12   .12   .13
14.     .30   .33   .30   .42   .43   .38   .35   .36   .33   .39   .34   .34
15.     .00   .00   .00   .00   .00   .00   .00   .00   .00   .00   .00   .00
16.    5.56  5.49  8.36  9.17 14.35 11.28 10.69  9.91 16.50  9.52  8.79  4.49
17.    7.72  6.81 10.08 11.38 21.26 15.70 13.66 15.01 27.48 10.97 13.82  6.96
18.    1.13   .86  1.22   .72  2.34  2.95  1.49  2.81  3.86   .38  2.72  2.63
19.      .297    .564    .085    .054
20.       77.     75.     77.     77.
21.       .15     .15     .15     .15
22.       .00     .00     .00     .00
23.     14.20   11.70    6.80   27.10   41.20
24.     .0038   .0038   .0076   .0114   .0019
25.      9.00   11.00   11.00   13.00
26.     10.00   10.00   10.00   10.00
27.      .050    .050    .050    .050    .050
28.      .050    .050    .050    .050    .050
29.       .00     .00     .00     .00
30.    750.00  750.00  750.00  750.00
31.      1.00    1.00    1.00    1.00
32.     55.00   55.00   50.00  100.00   60.00
33.     .0800   .0900   .0800   .0100   .0800
34.      9.00    9.00   11.00     .00
35.      2.00    2.00    2.50     .00
36.      .011    .011    .011    .000
37.     27.10   12.30    2.10     .00
38.      .050    .050    .050    .000
39.     10.00   10.00   10.00     .00
40.       .31     .31     .31     .00
41.     1.000   1.000   1.000    .000
42.       .19     .19     .19     .19
43.    141.00  267.00  102.00   26.00
44.     24.00   24.00   24.00   24.00
45.     20.00   20.00   20.00   20.00
46.    400.00  300.00  450.00  251.00
47.    400.00  300.00  450.00  250.00
48.       .08     .08     .08     .08
49.       .14     .29     .42     .00
50.    259.00  450.00  144.00     .00
51.     89.00   89.00   89.00     .00
52.     75.00  144.00   28.00     .00
53.     29.10   23.80   27.30     .00
54.     24.00   18.00   22.00     .00
55.       .14     .12     .12     .00
56.    200.00  200.00  200.00     .00
57.    250.00  250.00  250.00     .00
58.       .08     .08     .08     .00
59.     4
60.     1   2   3   4
61.     7     .31   81.00
62.     10.00  150.00  510.00  840.00 1190.00 1520.00 1900.00
63.      1.40    1.40    1.55    1.58    1.55    1.55    1.55
64.       .19     .19     .19     .19     .18     .18     .18
```

```
     16.00   16.00   15.00   12.00   12.00   12.00   12.00
     16.00   16.00   23.00   35.00   35.00   30.00   30.00
   1900.00
     7      .25   50.00
     10.00  150.00  560.00  760.00  970.00 1420.00 1830.00
      1.42    1.42    1.42    1.60    1.60    1.60    1.60
       .13     .13     .08     .11     .13     .09     .08
     50.00   50.00   50.00   21.00   21.00   21.00   21.00
     15.00   19.00   19.00   19.00   19.00    9.00    9.00
   1830.00
     7      .35   81.00
     10.00  280.00  330.00  530.00  760.00 1140.00 1905.00
      1.42    1.42    1.57    1.55    1.55    1.55    1.55
       .15     .15     .15     .14     .14     .14     .14
      7.00    5.00     .75     .75     .75     .75     .75
     23.00   23.00   34.00   45.00   44.00   42.00   42.00
   1905.00
     7      .31   84.00
     10.00  280.00  560.00  860.00 1270.00 1500.00 1828.00
      1.42    1.42    1.42    1.45    1.45    1.45    1.45
       .19     .19     .19     .18     .18     .18     .18
     12.00   12.00   12.00   12.00   12.00   12.00   12.00
     12.00   23.00   39.00   44.00   42.00   42.00   42.00
   1828.00
     3  15   9  15   1   1
      .170   4.000    .000
     0     .00     .00
     3  15   9  15   1   2
      .130   4.000    .000
     0     .00     .00
     3  15   9  15   1   3
      .100   4.000    .000
     0     .00     .00
     3  15   9  15   1   4
      .050   4.000    .000
     0     .00     .00
  70  1  0.0  0.0  0.0  0.0
  70  2  0.0  0.0  0.0  0.0
  70  3  0.0  0.0  0.0  0.0
  70  4  0.0  0.0  0.0  0.0
  70  5  0.0  0.0  0.0  0.0
  70  6  0.0  0.0  0.0  0.0
  70  7  0.0  0.0  0.0  0.0
  70  8  0.0  0.0  0.0  0.0
  70  9  0.0  0.0  0.0  0.0
  70 10  0.0  0.0  0.0  0.0
               .
               .
               .
               .
               .
```

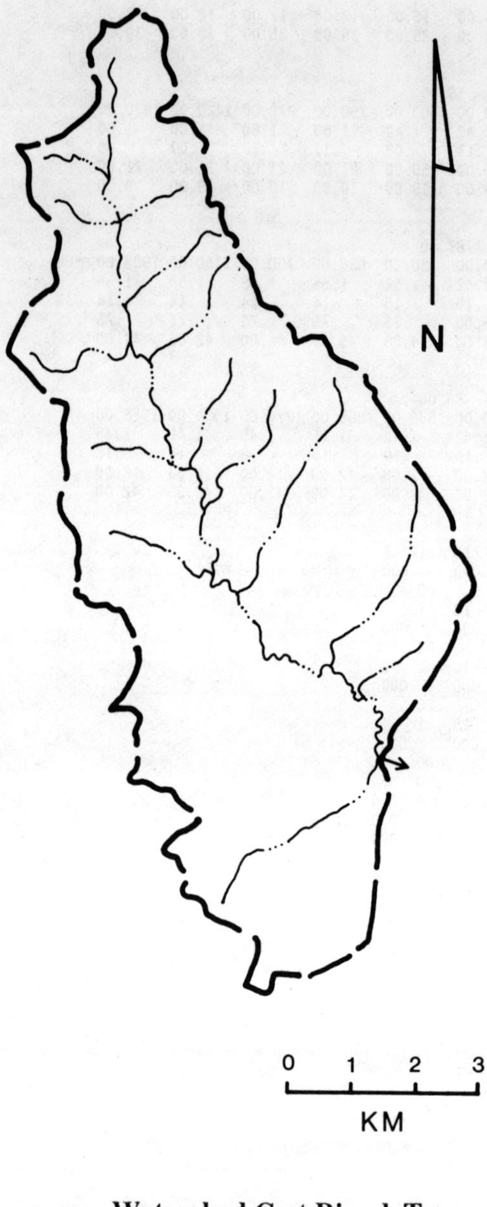

Watershed G at Riesel, Texas

*** SWRRB DATA ASSEMBLY FORM 1 ***

1. **TITLE:** Riesel, Texas
Watershed Station G
Brazos River Basin

2. **PROGRAM CONTROL CODES**

Number of years of runoff simulation ___6___

Beginning year of runoff simulation ___1961___

Number of subareas in basin ___1___

Printout frequency

Monthly (0) ___X___ Daily (1) ______ Annual (2) ______

Will the rainfall be:

Read in single rain gage for entire basin	(1)	X
Simulated single rain gage for entire basin	(2)	
Read in one rain gage for each sub-basin	(3)	
Simulated for multiple rain gages	(4)	

Will the max & min temperatures be:

Read in single max & min for entire basin	(1)	
Simulated single max & min for entire basin	(2)	X
Read in max & min for each sub-basin	(3)	
Simulated for each sub-basin	(4)	

Number of times random number generator cycles before simulation begins ___0___

Code for stat collection on monthly water yield

Skip collection	(0)	
Collect	(1)	X

Code for stat collection on monthly sediment yield

Skip collection	(0)	X
Collect	(1)	

*** SWRRB DATA ASSEMBLY FORM 2 ***

3. GENERAL DATA

Basin area (km^2) 17.73

Rainfall correction factor 1.0

Baseflow factor .05

Basin lag time (days) 10

Initial soil water storage -- fraction of field capacity 0

(entering 0. allows SWRRB to estimate it)

4. CENTROID COORDINATES OF SUB-BASINS

X: _____ _____ _____ _____ _____ Y: _____ _____ _____ _____ _____

5. GENERAL WEATHER DATA

If automatic entry is used, enter:

Weather station No. (Appendix I) _________ and skip to Section 10. If not, continue

TP-40 10-year frequency 0.5 hour rainfall (mm) _________

TP-40 10-year frequency 6.0 hour rainfall (mm) _________

Number of years for maximum monthly 0.5 hour rainfall record _________

Latitude of watershed (deg) _________

6A. MONTHLY TEMPERATURE DATA

Average monthly maximum air temperature (°C)

_____ _____ _____ _____ _____ _____ _____ _____ _____ _____ _____ _____

Average monthly minimum air temperature (°C)

_____ _____ _____ _____ _____ _____ _____ _____ _____ _____ _____ _____

Coefficient of variation for monthly temperature

_____ _____ _____ _____ _____ _____ _____ _____ _____ _____ _____ _____

*** SWRRB DATA ASSEMBLY FORM 4 ***

10. BASIN DATA

Sub-basin:	1	2	3	4	5	Entire basin
Fraction of basin in each subarea	1					
SCS runoff curve number - Condition II	84					
Soil albedo	.10					
Water content of snow on ground at start of simulation (mm)	0					
Main channel length in each sub-basin (km) of simulation (mm)	7.8					7.8
Average channel slope in each sub-basin (m/m)	.0034					.0034
Average main channel width (m)	6.1					
Effective hydraulic conductivity of channel alluvium (mm/hr)	.001					
Channel N value	.05					.05
Overland flow N value	.05					.05
Return flow travel time (days) (skip if unknown)	0					
Sediment concentration in return flow (ppm)	500					
USLE erosion control practice factor P	.50					
Average slope length for each sub-basin (m)	50					50
Average slope steepness for each sub-basin (m/m)	.021					.021

11. ROUTING DATA (If only one sub-basin is used, skip to Section 12)

(All routing data is from sub-basin to basin outlet)

	1	2	3	4	5
Average channel width (m)					
Average channel depth (m)					
Channel slope (m/m)					
Channel length (km)					
Channel N value (mm/hr)					
Effective hydraulic conductivity of channel alluvium (mm/hr)					
USLE soils K factor for channel					
USLE C factor for channel					

*** SWRRB DATA ASSEMBLY FORM 5 ***

12. POND DATA

(If not used skip to Section 13)

Sub-basin:	1	2	3	4	5
Fraction of each sub-basin that flows into ponds	.2				
Total surface area of all ponds in each sub-basin (ha)	17.5				
Runoff volume from pond catchment area to fill ponds (mm)	24				
Initial pond volumes (mm)	20				
Initial sediment concentration in ponds (ppm)	350				
Normal sediment concentration in ponds (ppm)	200				
Hydraulic conductivity of pond bottoms (mm/hr)	.08				

13. RESERVOIR DATA (If not used skip to Section 14)

Fraction of each sub-basin that flows in reservoirs					
Total reservoir surface area at emergency spillway (ha)					
Runoff volume from reservoir catchment area required to fill emergency spillway (mm)					
Total reservoir surface area at principal spillway (ha)					
Runoff required to fill to principal spillway (mm)					
Initial reservoir volumes (mm)					
Average principal spillway release rate ($m^3/s/km^2$)					
Initial sediment concentration in reservoirs (ppm)					
Normal sediment concentration in reservoirs (ppm)					
Hydraulic conductivity of reservoir bottoms (mm/hr)					

*** SWRRB DATA ASSEMBLY FORM 6 ***

14. SOILS DATA

Number of soils 1

Soil Series name

#1 Houston black #2 ______ #3 ______ #4 ______ #5 ______

Subbasin:	1	2	3	4	5
Soil Series for each subbasin:	1				
No. layers for each soil series (if automatic entry is used, skip to Section 15)					

Soil Series name ______ K Factor ______ Pass #200 Sieve ______

Layer:

	1	2	3	4	5	6	7	8	9	10
Depth to bottom layer (mm)										
Bulk density (m/m^3)										
Available water capacity (mm/mm)										
Saturated conductivity (mm/h)										
Clay content (%)										

Sediment size distribution (fraction-if unknown, skip to next soil)

Sand (.20 mm)	Silt (.01 mm)	Clay (.002 mm)	Sm Ag (.03 mm)	L Ag (.50 mm)

Maximum rooting depth (mm) ______

*** SWRRB DATA ASSEMBLY FORM 7 ***

Subbasin 1

15. CROP DATA

Planting date: Month 2 Day 1

Harvest date: Month 9 Day 10

(Use start and end of growing season for perennial vegetation)

Vegetation: Annual (1) X Perennial (2) ____

Tillage Operation:

- Fall plow (1) ____
- Spring plow (2) ____
- Conservation tillage (3) X
- Zero tillage (4) ____

Average annual C-factor .15

Maximum leaf area index 5

Initial residue cover (kg/ha) ____

16. IRRIGATION DATA

Are the crops irrigated Yes (1) ____ No (0) X

Water stress when irrigation is applied .9

Irrigation runoff ratio 1

17. DAILY RAINFALL AND TEMPERATURE DATA

If rainfall is input, list daily rainfall (mm) for each subbasin.

If temperature is input, list maximum and minimum temperature for each subbasin in a separate file.

[Format (5X, 15F5.1)]

```
RIESEL, TEXAS
WATERSHED STATION G
BRAZOS RIVER BASIN
   61961   1   0   1   2   0   0   0
  17.730   1.000    .050  10.000    .000
   85.09  181.61   26.00   31.37
   14.6  16.6  20.6  25.2  29.3  33.6  35.6  35.9  32.3  27.3  20.1  16.0
    3.2   5.0   8.2  13.2  17.7  21.9  23.7  23.5  19.9  14.2   7.4   4.3
    .17   .12   .15   .09   .06   .06   .04   .08   .06   .08   .11   .15
   250.  320.  427.  488.  562.  651.  613.  593.  503.  403.  306.  245.
  10.67 17.02 17.53 23.88 40.13 22.86 50.80 29.21 31.75 36.83  8.89 12.19
    .15   .21   .17   .20   .19   .14   .07   .11   .14   .12   .14   .13
    .40   .42   .42   .41   .43   .42   .34   .39   .46   .34   .43   .41
   6.86  7.37  7.62 12.19 15.75 11.68  8.13  9.91 13.97 13.72 10.41  8.13
   9.91  9.40  9.65 16.76 21.84 14.99 12.45 14.48 16.51 17.02 14.22 10.67
   2.36  2.25  2.01  2.19  2.26  2.10  2.79  3.77  1.97  1.86  3.12  2.25
   1.000
     84.
     .10
     .00
    7.80    7.80
   .0034   .0034
    6.10
     .00
     .05     .05
     .05     .05
     .00
  500.00
     .50
   50.00   50.00
   .0210   .0210
     .20
   17.50
   24.00
   20.00
  350.00
  200.00
     .08
   1
   1
   3      .32   92.00
   10.00  609.60 2032.00     .00     .00     .00     .00
    1.40    1.40    1.50     .00     .00     .00     .00
```

```
65.          .17     .17     .17     .00      .00      .00      .00
66.         1.52    1.52    1.52     .00      .00      .00      .00
67.        50.00   50.00   50.00     .00      .00      .00      .00
68.
69.      2032.00
70.         2   1    9  10    1   3
71.         .150    5.000
72.        1      .90    1.00
73.      1961    0.0
74.              0.0
75.              0.0
76.              0.0
77.               .
78.               .
79.               .
80.               .
```

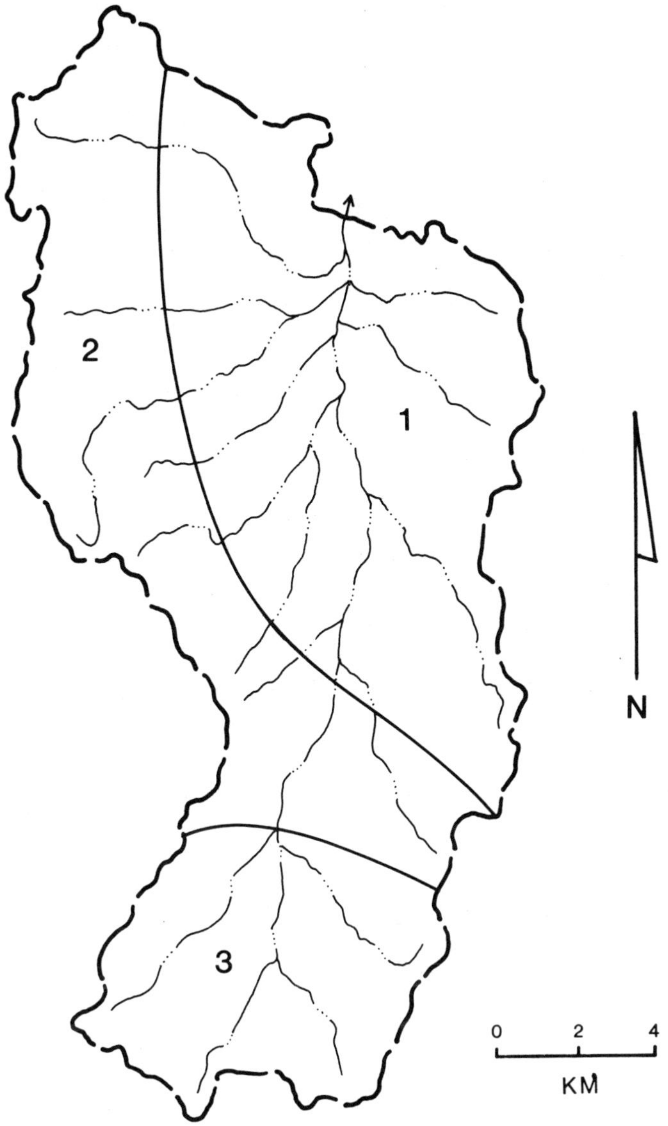

Watershed W-1 at Reynolds, Idaho

*** SWRRB DATA ASSEMBLY FORM 1 ***

1. TITLE: Reynolds, Idaho
Watershed W-1
Reynolds Creek Outlet

2. PROGRAM CONTROL CODES

Number of years of runoff simulation 2

Beginning year of runoff simulation 1975

Number of subareas in basin 3

Printout frequency

Monthly (0) X Daily (1) ____ Annual (2) ____

Will the rainfall be:

Read in single rain gage for entire basin (1) ____

Simulated single rain gage for entire basin (2) ____

Read in one rain gage for each sub-basin (3) X

Simulated for multiple rain gages (4) ____

Will the max & min temperatures be:

Read in single max & min for entire basin (1) ____

Simulated single max & min for entire basin (2) ____

Read in max & min for each sub-basin (3) X

Simulated for each sub-basin (4) ____

Number of times random number generator cycles before simulation begins 0

Code for stat collection on monthly water yield

Skip collection (0) X

Collect (1) ____

Code for stat collection on monthly sediment yield

Skip collection (0) X

Collect (1) ____

*** SWRRB DATA ASSEMBLY FORM 2 ***

3. GENERAL DATA

Basin area (km^2) 233.5

Rainfall correction factor 1.0

Baseflow factor .75

Basin lag time (days) 30

Initial soil water storage -- fraction of field capacity 0

(entering 0. allows SWRRB to estimate it)

4. CENTROID COORDINATES OF SUB-BASINS

X: ____ ____ ____ ____ ____ Y: ____ ____ ____ ____ ____

5. GENERAL WEATHER DATA

If automatic entry is used, enter:

Weather station No. (Appendix I) ______ and skip to Section 10. If not, continue

TP-40 10-year frequency 0.5 hour rainfall (mm) 19.5

TP-40 10-year frequency 6.0 hour rainfall (mm) 44.45

Number of years for maximum monthly 0.5 hour rainfall record 8.0

Latitude of watershed (deg) 43.5

6A. MONTHLY TEMPERATURE DATA

Average monthly maximum air temperature (°C)

2.28 5.78 10.83 17.39 22.17 26.94 33.00 31.50 26.00 18.50 9.11 4.11

Average monthly minimum air temperature (°C)

-5.5 -3.5 -0.05 3.00 6.94 10.61 15.00 13.05 8.11 3.28 -1.83 -3.89

Coefficient of variation for monthly temperature

.37 .29 .20 .15 .12 .12 .08 .15 .11 .14 .24 .26

*** SWRRB DATA ASSEMBLY FORM 3 ***

6B. MONTHLY TEMPERATURE DATA FOR REMAINING SUB-BASINS

(SKIP UNLESS TEMPERATURE IS SIMULATED FOR EACH SUB-BASIN)

2 Maximum (°C)

3

4

5

2 Minimum (°C)

3

4

5

7. MONTHLY SOLAR RADIATION DATA

Monthly average daily solar radiation (ly)

138 236 342 485 585 636 670 576 460 301 182 124

8. MONTHLY RAINFALL DATA

Monthly maximum 0.5 h rainfall (mm)

8.1 11.4 28.4 25.1 42.4 55.4 39.1 28.4 31.7 22.9 26.7 13.5

9. PRECIPITATION GENERATION PARAMETER TABLE

(Enter P(W/D) & P(W/W) or DAYPR)

P(W/D)

.32 .23 .22 .21 .20 .15 .05 .06 .08 .15 .21 .27

P(W/W)

.59 .56 .46 .41 .48 .46 .25 .35 .37 .39 .53 .54

Days Prec

Mean

3.30 2.79 2.54 3.81 4.06 3.81 2.29 3.56 3.30 3.05 3.56 2.79

St. dev.

4.06 3.30 3.05 4.83 5.08 5.33 4.06 4.83 3.81 2.79 3.56 3.81

Skew coefficient

2.68 2.82 3.10 2.73 2.14 4.87 4.22 1.86 1.91 1.30 1.42 2.80

*** SWRRB DATA ASSEMBLY FORM 4 ***

10. BASIN DATA

Sub-basin:	1	2	3	4	5	Entire basin
Fraction of basin in each subarea	.381	.507	.112			
SCS runoff curve number - Condition II	56	72	65			
Soil albedo	.15	.15	.15			
Water content of snow on ground at start of simulation (mm)	10	20	60			
Main channel length in each sub-basin (km) of simulation (mm)	11.6	6	3.50			22.2
Average channel slope in each sub-basin (m/m)	.0227	.0730	.1234			.0412
Average main channel width (m)	9	3	3			
Effective hydraulic conductivity of channel alluvium (mm/hr)	10	10	10			
Channel N value	.10	.10	.10			.10
Overland flow N value	.05	.05	.05			.05
Return flow travel time (days) (skip if unknown)	0	0	0			
Sediment concentration in return flow (ppm)	500	500	500			
USLE erosion control practice factor P	1	1	1			
Average slope length for each sub-basin (m)	80	45	75			50
Average slope steepness for each sub-basin (m/m)	.05	.10	.12			.09

11. ROUTING DATA (If only one sub-basin is used, skip to Section 12)

(All routing data is from sub-basin to basin outlet)

	1	2	3	4	5
Average channel width (m)	0	9	9		
Average channel depth (m)	0	3	2.5		
Channel slope (m/m)	0	.033	.014		
Channel length (km)	0	5.8	15.6		
Channel N value (mm/hr)	0	.10	.10		
Effective hydraulic conductivity of channel alluvium (mm/hr)	0	10	10		
USLE soils K factor for channel	0	.24	.24		
USLE C factor for channel	0	1	1		

*** SWRRB DATA ASSEMBLY FORM 6 ***

14. SOILS DATA

Number of soils 2

Soil Series name

#1 Nannyton #2 Harmehl #3 ____ #4 ____ #5 ____

Subbasin:	1	2	3	4	5
Soil Series for each subbasin:	1	1	2		
No. layers for each soil series (if automatic entry is used, skip to Section 15)	4	4			

Soil Series name Nannyton K Factor .04 Pass #200 Sieve 52

Layer:

	1	2	3	4	5	6	7	8	9	10
Depth to bottom layer (mm)	10	228.6	771	1524						
Bulk density (m/m^3)	1.45	1.45	1.50	1.55						
Available water capacity (mm/mm)	0.14	0.14	0.13	0.04						
Saturated conductivity (mm/h)	15	15	15	35						
Clay content (%)	20	20	28	20						

Sediment size distribution (fraction-if unknown, skip to next soil)

Sand (.20 mm)	Silt (.01 mm)	Clay (.002 mm)	Sm Ag (.03 mm)	L Ag (.50 mm)

Maximum rooting depth (mm) ____________

*** SWRRB DATA ASSEMBLY FORM 6 ***

14. SOILS DATA

Number of soils ___________

Soil Series name

#1 ___________ #2 ___________ #3 ___________ #4 ___________ #5 ___________

Subbasin:	1	2	3	4	5
Soil Series for each subbasin:	___	___	___	___	___
No. layers for each soil series (if automatic entry is used, skip to Section 15)	___	___	___	___	___

Soil Series name Harmehl K Factor .04 Pass #200 Sieve 62

Layer:

	1	2	3	4	5	6	7	8	9	10
Depth to bottom layer (mm)	10	432	711	991						
Bulk density (m/m^3)	1.45	1.45	1.50	1.55						
Available water capacity (mm/mm)	0.13	0.13	0.13	0.13						
Saturated conductivity (mm/h)	15	15	15	15						
Clay content (%)	16	16	16	20						

Sediment size distribution (fraction-if unknown, skip to next soil)

Sand (.20 mm)	Silt (.01 mm)	Clay (.002 mm)	Sm Ag (.03 mm)	L Ag (.50 mm)
___	___	___	___	___

Maximum rooting depth (mm) ___________

*** SWRRB DATA ASSEMBLY FORM 7 ***

Subbasin ____1____

15. CROP DATA

Planting date: Month ___3___ Day ___15___

Harvest date: Month ___10___ Day ___10___

(Use start and end of growing season for perennial vegetation)

Vegetation: Annual (1) ________ Perennial (2) ___X___

Tillage Operation:

- Fall plow (1) ________
- Spring plow (2) ________
- Conservation tillage (3) ________
- Zero tillage (4) ___X___

Average annual C-factor ___.10___

Maximum leaf area index ___4.0___

Initial residue cover (kg/ha) ________

16. IRRIGATION DATA

Are the crops irrigated Yes (1) ________ No (0) ___X___

Water stress when irrigation is applied ________

Irrigation runoff ratio ________

17. DAILY RAINFALL AND TEMPERATURE DATA

If rainfall is input, list daily rainfall (mm) for each subbasin.

If temperature is input, list maximum and minimum temperature for each subbasin in a separate file.

[Format (5X, 15F5.1)]

*** SWRRB DATA ASSEMBLY FORM 7 ***

Subbasin __2__

15. CROP DATA

Planting date: Month __4__ Day __1__

Harvest date: Month __10__ Day __1__

(Use start and end of growing season for perennial vegetation)

Vegetation: Annual (1) ____ Perennial (2) __X__

Tillage Operation:

- Fall plow (1) ____
- Spring plow (2) ____
- Conservation tillage (3) ____
- Zero tillage (4) __X__

Average annual C-factor __.05__

Maximum leaf area index __4.0__

Initial residue cover (kg/ha) ____

16. IRRIGATION DATA

Are the crops irrigated Yes (1) ____ No (0) __X__

Water stress when irrigation is applied ____

Irrigation runoff ratio ____

17. DAILY RAINFALL AND TEMPERATURE DATA

If rainfall is input, list daily rainfall (mm) for each subbasin.

If temperature is input, list maximum and minimum temperature for each subbasin in a separate file.

[Format (5X, 15F5.1)]

*** SWRRB DATA ASSEMBLY FORM 7 ***

Subbasin 3

15. CROP DATA

Planting date: Month 4 Day 15

Harvest date: Month 9 Day 20

(Use start and end of growing season for perennial vegetation)

Vegetation: Annual (1) ____ Perennial (2) X

Tillage Operation:

- Fall plow (1) ____
- Spring plow (2) ____
- Conservation tillage (3) ____
- Zero tillage (4) X

Average annual C-factor .05

Maximum leaf area index 4.0

Initial residue cover (kg/ha) ____

16. IRRIGATION DATA

Are the crops irrigated Yes (1) ____ No (0) X

Water stress when irrigation is applied ____

Irrigation runoff ratio ____

17. DAILY RAINFALL AND TEMPERATURE DATA

If rainfall is input, list daily rainfall (mm) for each subbasin.

If temperature is input, list maximum and minimum temperature for each subbasin in a separate file.

[Format (5X, 15F5.1)]

```
REYNOLDS, IDAHO
WATERSHED W-1
REYNOLDS CREEK OUTLET
  021975   3   0   1   3   0
  233.50    1.00    0.75    30.0    0.00
   19.50   44.45    8.00    43.5
   2.28  5.78 10.83 17.39 22.17 26.94 33.00 31.50 26.00 18.50  9.11  4.11
  -5.50 -3.05 -0.05  3.00  6.94 10.61 15.00 13.05  8.11  3.28 -1.83 -3.89
   0.37  0.29  0.20  0.15  0.12  0.12  0.08  0.15  0.11  0.14  0.24  0.26
  138.0 236.0 342.0 485.0 585.0 636.0 670.0 576.0 460.0 301.0 182.0 124.0
    8.1  11.4  28.4  25.1  42.4  55.4  39.1  28.4  31.7  22.9  26.7  13.5
  0.317 0.235 0.223 0.211 0.196 0.150 0.053 0.063 0.083 0.152 0.213 0.271
  0.595 0.559 0.459 0.406 0.476 0.464 0.250 0.353 0.370 0.389 0.534 0.543
  3.302 2.794 2.540 3.810 4.064 3.810 2.286 3.556 3.302 3.048 3.556 2.790
  4.064 3.302 3.048 4.826 5.080 5.334 4.064 4.826 3.810 2.794 3.556 3.810
  2.680 2.820 3.100 2.730 2.140 4.870 4.220 1.860 1.910 1.300 1.420 2.800
    0.381   0.507   0.112
     56.     70.     60.
     0.15    0.15    0.15
    10.00   20.00   60.00
    11.60    6.00    3.50   22.20
   0.0227  0.0730  0.1234  0.0412
     9.00    3.00    3.00
    10.00   10.00   10.00
     0.10    0.10    0.10    0.10
     0.05    0.05    0.05    0.05
     0.00    0.00    0.00
   500.00  500.00  500.00
     1.00    1.00    1.00
    80.00   45.00   75.00   50.00
     0.05    0.10    0.12    0.09
     0.00    9.00    9.00
     0.00    3.00    3.00
     0.00   .0330   0.024
     0.00     5.8    15.6
     0.00    0.10    0.10
     0.00   10.00   10.00
     0.00    0.24    0.24
     0.00    1.00    1.00
    2
    4   4
    1   1   2
     0.04     52.0
    10.00  228.60  711.00 1524.00
     1.45    1.45    1.50    1.55
     0.14    0.14    0.13    0.04
    15.00   15.00   15.00   15.00
```

```
        20.00   20.00   28.00   20.00
         0.04   62.0
        10.00  432.00  711.00  991.00
         1.45    1.45    1.50    1.55
         0.13    0.13    0.13    0.13
        15.00   15.00   15.00   15.00
        16.00   16.00   16.00   20.00
         3  15  10  10   2   4
         0.10    4.0
         4   1  10   1   2   4
         0.05    4.0
         4  15   9  20   2   4
         0.05    4.0
       1967  0.0  0.0  0.0
       1967  0.0  0.0  0.0
       1967  0.5  0.6  7.3
       1967  0.5  1.0  5.2
       1967  0.0  2.3  4.0
       1967  0.0  1.6 10.5
       1967  0.0  0.0  0.0
       1967  0.0  0.0  0.0
       1967  0.0  0.0  0.0
       1967  0.0  0.0  0.0
                  .
                  .
                  .
                  .
                  .
                  .
                  .
```

TEMPERATURE FILE FOR REYNOLDS, IDAHO

1.	1975	-5.	-6.	-4.	-16.	-8.	-12.
2.	1975	1.	-3.	-8.	-11.	-9.	-12.
3.	1975	3.	3.	-3.	-8.	-3.	-10.
4.	1975	4.	2.	-2.	-1.	-3.	-8.
5.	1975	4.	2.	-2.	-6.	-4.	-8.
6.	1975	4.	2.	-2.	0.	-2.	-6.
7.	1975	5.	3.	-1.	-1.	-3.	-6.
8.	1975	6.	3.	-1.	-4.	-6.	-10.
9.	1975	-2.	-7.	-11.	-8.	-11.	-15.
10.	1975	0.	-3.	-6.	-5.	-7.	-12.
11.	1975	-3.	-6.	-9.	-11.	-9.	-12.
12.	1975	-1.	0.	-3.	-9.	-8.	-9.
13.	1975	-1.	6.	-1.	-10.	-2.	-4.
14.	1975	-7.	7.	1.	1.	2.	-2.
15.	1975	5.	6.	3.	-2.	0.	-1.
16.	1975	7.	3.	0.	-2.	0.	-3.
17.	1975	11.	7.	3.	0.	2.	-2.
18.	1975	10.	7.	2.	0.	2.	-3.
19.	1975	7.	7.	6.	-3.	1.	-3.
20.	1975	8.	6.	1.	-1.	-2.	-7.
21.	1975	2.	-1.	-1.	-6.	-6.	-11.
22.	1975	1.	4.	2.	-7.	-5.	-8.
23.	1975	8.	6.	1.	-5.	-1.	-1.
24.	1975	11.	7.	2.	1.	3.	0.
25.	1975	12.	8.	3.	6.	4.	0.
26.	1975	9.	5.	1.	-7.	-9.	-12.
27.	1975	-3.	-6.	-10.	-9.	-11.	-15.
28.	1975	-1.	-5.	-10.	-10.	-10.	-15.
29.	1975	-2.	-6.	-8.	-12.	-11.	-17.
30.	1975	-4.	-8.	-13.	-15.	-12.	-17.
31.	1975	-1.	-2.	-7.	-13.	-8.	-12.
32.	1975	6.	2.	-3.	-2.	-5.	-7.
33.	1975	5.	2.	-3.	-1.	-1.	-7.
34.	1975	4.	0.	-4.	-2.	-3.	-7.
35.	1975	1.	-2.	-5.	-1.	-4.	-9.
36.	1975	3.	-1.	-4.	-4.	-6.	-10.
37.	1975	1.	1.	-1.	-3.	-4.	-8.
38.	1975	3.	5.	0.	-2.	1.	-3.
39.	1975	4.	2.	0.	-1.	0.	-3.
40.	1975	8.	4.	1.	-2.	0.	-3.
41.	1975	6.	1.	-2.	0.	-1.	-5.
42.	1975	6.	3.	-1.	1.	-1.	-6.
43.	1975	8.	4.	-1.	3.	1.	-2.
44.	1975	9.	4.	0.	2.	1.	-3.
45.	1975	2.	1.	-4.	-2.	-6.	-11.
46.	1975	2.	0.	-3.	-6.	-8.	-11.
47.	1975	0.	-2.	-8.	-3.	-7.	-11.
48.	1975	1.	-2.	-8.	-6.	-8.	-11.
49.	1975	4.	1.	-4.	-6.	-4.	-10.
50.	1975	4.	2.	-1.	-1.	-1.	-6.
51.	1975	2.	-1.	-5.	-6.	-9.	-13.
52.	1975	-1.	-6.	-11.	-7.	-10.	-14.
53.	1975	4.	1.	-6.	-5.	-7.	-13.
54.				.			
55.				.			
56.				.			
57.				.			
58.				.			

Output Interpretation

A listing of the code names of the output variables and their definitions are given in Appendix X so users can modify code and tailor the output to suit their particular needs. Here the existing output is interpreted for the simulation run from Chickasha, Oklahoma. The output is divided into pages and described as follows:

Page 1

Page 1 contains a listing of several input variables for inspection. Also included are the random number generator seeds at the start of simulation.

Page 2

Page 2 begins by notifying the user which rainfall and temperature input codes were selected. Monthly rainfall generator parameters were input by the user and include:

W|D = Probability of a wet day following a dry day

W|W = Probability of a wet day following a wet day

MEAN = Mean daily rainfall amount in mm

ST DV = Standard deviation of daily rainfall amounts in mm

SKW CF = Skew coefficient of daily rainfall amounts

The last section of Page 2 includes several input and calculated weather variables including:

R5MX = Maximum 0.5 h rainfall in mm

TMX = Average maximum temperature in °C

TMN = Average minimum temperature in °C

RA = Average solar radiation in ly

CVT = Coefficient of variation for temperature

RAIN = Average rainfall in mm

DAYP = Average number of days of precipitation

ALPH = Average $\alpha .5$ (maximum 0.5 h rainfall/total event rainfall)

Pages 3, 4, and 5

These pages include basin hydrology and sedimentation inputs. Also included are pond and reservoir inputs and routing data. The output is simply a printing of the inputs for inspection and the descriptions are self-explanatory.

Page 6

Soils data is listed for each subbasin starting with subbasin number one through subbasin number four. Soils data includes depth to the bottom of the layer, porosity, 15 bar (assumed wilting point) soil water content, 0.3 bar (assumed field capacity) soil water content, available water content, initial water content, and saturated conductivity. The data has been input or directly calculated from the soils input. Total available soil water (.3 bar-15 bar) and the initial total profile soil water (soil water-15 bar) are shown for each subbasin. The initial composite state is the average soil water content weighted by subbasin area.

Page 7

Page 7 contains soils data that has been input for the surface soil layer including the percent sand, silt, and clay and the USLE K-factor. Page 7 also shows the sediment size distribution of the detached sediment for each subbasin.

Page 8

Crop data that has been input by the user is printed for each subbasin on the top of Page 8. Potential heat units (°C) are computed from the planting and harvest dates. Maximum plant rooting depth is input or set to the depth of the soil profile by default.

The irrigation data on Page 8 includes the water stress that triggers an irrigation and the runoff ratio. These values have been input by the user.

Page 9

Page 9 contains precipitation, surface runoff, subsurface flow, water yield, percolation, transmission losses, ET, sediment yield, and soil water content. The values are basin composite values and are weighted by subbasin area. They are printed by month and summed for the entire year. Soil water values are printed at the end of each month.

Page 10

Soil water values at the end of the simulation are printed on the top of Page 10. Following the soil water values is a listing of pond and reservoir water volumes and sediment concentrations at the end of the simulation. The final composite pond and reservoir storages are basin values weighted by subbasin area. Average annual irrigation data for each subbasin, including number of applications and irrigation water applied, are printed for user inspection. The bottom of Page 10 includes soil, pond, and reservoir water balance and pond and reservoir sediment balance. Values significantly different than zero may indicate unaccounted water and sediment entering or exiting the system.

Page 11

Subbasin average annual values for rainfall, surface runoff, subsurface flow, sediment yield, and total biomass is shown on the top of Page 11. Next, average monthly basin values for rainfall, snowfall, surface runoff, subsurface flow, water yield, ET, and sediment yield area listed. Finally, selected miscellaneous basin statistics are shown. Standard deviation of rainfall are in mm. The mean CN is for the entire basin while the maximum and minimum are for individual subbasins. Basin peak flow statistics are listed along with the mean and standard deviation of monthly basin water yields.

Page 12

Average annual basin values are weighted by subbasin areas and most definitions and units are self-explanatory. Total subbasin sediment yield is the sum of the subbasin yields. Basin sediment yield is the total sediment yield reaching the basin outlet. The units of the variables in the pond and reservoir water budget are in mm over the area draining into the ponds and reservoirs. The yield loss from ponds and reservoirs is the amount of water and sediment the ponds and reservoirs trapped from going downstream.

Page 13

Measured and predicted water yields are compared on Page 13. Monthly error is absolute error while annual error is percent error. The data set was developed to demonstrate several input options and not describe the system. Consequently, measured and predicted values may not compare well.

Page 14

Monthly measured and predicted means and standard deviations, regression line slope, and R^2 values are shown on top of Page 14. Next, average monthly measured and predicted water yields are listed to determine if the model is overpredicting or underpredicting certain seasons of the year.

Pages 15 and 16

These pages show an identical statistical analysis for sediment yields as was performed for water yields on Pages 13 and 14.

```
SWRRB    1/01/89 IBM PC  VERSION

CHICKASHA, OKLAHOMA
WATERSHED 522
LITTLE WASHITA RIVER BASIN
     NO YRS =    3
     BASIN AREA =   538.200 KM**2
     AVE A RAINFALL/AVE A FOR GAGE = 1.000
     BASEFLOW FACTOR =      .500
     BASIN LAG TIME =   30.00 D
     GENERATOR CYCLES =    0
     WATER STATS =  1
     SEDIMENT STATS =  1


     GENERATOR SEEDS
             9  98 915  92
           135  28 203  85
            43  54 619  33
           645   9 948  65
           885  41 696  62
            51  78 648   0
           227  57 929  37
            20  90 215  31
           320  73 631  49

     CENTROID COORDINATES OF SUB AREAS(KM)
     9.34   12.29   22.11   12.29   28.18   19.90   24.21   11.03

     TP-40 RAINFALL AMOUNTS (10 YR FREQ) FOR DUR
          0.5 H =  56.00 MM
          6H=108.00 MM

     NO YRS RECORD MAX.5H RAIN=  8.0

     LATITUDE= 35.10 DEG
```

SWRRB 1/01/89 IBM PC VERSION

CHICKASHA, OKLAHOMA
WATERSHED 522
LITTLE WASHITA RIVER BASIN
CLIMATE DATA

RAINFALL DATA USED IN THIS RUN ARE:
MEASURED MULTIPLE RAINGAGES

TEMPERATURE DATA USED IN THIS RUN ARE;
SIMULATED SINGLE TEMP FOR ENTIRE BASIN

-MO RAIN PROB--		-MO STATS FOR DAILY RAIN-		
W/D	W/W	MEAN	ST DV	SKW CF
.080	.300	5.560	7.720	1.130
.120	.330	5.490	6.810	.860
.130	.300	8.360	10.080	1.220
.160	.420	9.170	11.380	.720
.180	.430	14.350	21.260	2.340
.220	.380	11.280	15.700	2.950
.120	.350	10.690	13.660	1.490
.180	.360	9.910	15.010	2.810
.200	.330	16.500	27.480	3.860
.120	.390	9.520	10.970	.380
.120	.340	8.790	13.820	2.720
.130	.340	4.490	6.960	2.630

	R5MX	TMX	TMN	RA	CVT	RAIN	DAYP	ALPH
JAN	8.10	8.94	-3.48	220.00	.34	17.68	3.18	.40
FEB	11.40	11.60	-1.78	291.00	.24	24.18	4.41	.50
MAR	28.40	16.90	3.23	370.00	.22	40.59	4.86	.45
APR	25.10	23.40	9.77	453.00	.15	59.48	6.49	.53
MAY	42.40	27.00	14.01	511.00	.12	106.76	7.44	.45
JUN	55.40	31.60	19.30	554.00	.07	88.63	7.86	.57
JUL	39.10	34.30	21.70	544.00	.06	51.65	4.83	.59
AUG	28.40	32.80	20.30	501.00	.06	67.44	6.80	.51
SEP	31.70	28.40	16.90	392.00	.11	113.79	6.90	.30
OCT	22.90	23.60	9.82	332.00	.14	48.51	5.10	.48
NOV	26.70	16.50	3.58	237.00	.19	40.57	4.62	.43
DEC	13.50	10.80	-1.46	197.00	.27	22.90	5.10	.56
YR	27.76	22.15	9.32	383.50	.16	682.19	67.57	.48

PAGE 2

```
SWRRB    1/01/89 IBM PC  VERSION

CHICKASHA, OKLAHOMA
WATERSHED 522
LITTLE WASHITA RIVER BASIN
SUB-BASIN DATA


     SUB-BASIN AREA/BASIN AREA
  .297      .564      .085      .054

     POND CATCHMENT AREA FRACTION
  .190      .190      .190      .190

     POND SURFACE AREA(HA)
141.00    267.00    102.00     26.00

     MAX POND STORAGE(MM)
  24.0      24.0      24.0      24.0

     INTIAL POND STORAGE(MM)
  20.0      20.0      20.0      20.0

     INITIAL SED CONC IN PONDS(PPM)
  400.      300.      450.      251.

     NORMAL SED CONC IN PONDS(PPM)
  400.      300.      450.      250.

     SAT CONDUCTIVITY FOR POND BOTTOMS(MM/H)
   .08       .08       .08       .08

     RESERVOIR CATCHMENT AREA FRACTION
  .140      .290      .420      .000

     RESERVOIR SURFACE AREA AT EMERGENCY SPILLWAY(HA)
259.00    450.00    144.00       .00

     RESERVOIR STORAGE AT EMERGENCY SPILLWAY(MM)
  89.0      89.0      89.0        .0

     RESERVOIR SURFACE AREA AT PRINCIPAL SPILLWAY(HA)
 75.00    144.00     28.00       .00

     RESERVOIR STORAGE AT PRINCIPAL SPILLWAY(MM)
  29.1      23.8      27.3        .0

     INITIAL RESERVOIR STORAGE(MM)
  24.0      18.0      22.0        .0

     AVE RESERVOIR RELEASE RATES(M**3/S/KM**2)
.14000    .12000    .12000    .00000
```

```
     INITIAL SED CONC IN RESERVOIRS (PPM)
  200.      200.      200.       0.

     NORMAL SED CONC IN RESERVOIRS (PPM)
  250.      250.      250.       0.

     SAT CONDUCTIVITY OF RESERVOIR BOTTOMS (MM/H)
   .08       .08       .08       .00

     2 COND CN
  77.0      75.0      77.0      77.0

     SOIL ALBEDO
   .15       .15       .15       .15

     WATER CONTENT OF SNOW COVER (MM)
    .0        .0        .0        .0

     MAIN CHANNEL LENGTH (KM)
 14.20     11.70      6.80     27.10     41.20

     CHANNEL SLOPE(M/M)
 .0038     .0038     .0076     .0114     .0019

     AVERAGE MAIN CHANNEL WIDTH (M)
  9.00     11.00     11.00     13.00

     HYDR COND OF CHANNEL ALLUVIUM(MM/H)
 10.00     10.00     10.00     10.00

     CHANNEL N VALUE
  .050      .050      .050      .050      .050

     OVERLAND FLOW N VALUE
  .050      .050      .050      .050      .050

     TIME OF CONCENTRATION FOR SUB-BASINS(H)
  2.21      1.73      1.07      3.70      6.68

     RET FLO SED CONC (PPM)
  750.      750.      750.      750.

     RET FLO TRAVEL TIME(D)
  .000      .000      .000      .000

     SLOPE LENGTH(M)
   55.       55.       50.      100.       60.

     SLOPE STEEPNESS(M/M)
 .0800     .0900     .0800     .0100     .0800

     EROSION CONTROL PRACTICE FACTORS(P)
  1.00      1.00      1.00      1.00

     SLOPE LENGTH AND STEEPNESS FACTORS(LS)
  1.42      1.70      1.35       .15
```

```
          ROUTING DATA -- SUB-BASIN TO BASIN OUTLET

    AVE CHANNEL WIDTH(M)
 9.00       9.00     11.00        .00

    AVE CHANNEL DEPTH(M)
 2.00       2.00      2.50        .00

    CHANNEL SLOPE(M/M)
.0110      .0110     .0110      .0000

    CHANNEL LENGTH(KM)
27.10      12.30      2.10        .00

    CHANNEL N VALUE
  .05        .05       .05        .00

    HYDR COND OF CHANNEL ALLUVIUM(MM/H)
10.00      10.00     10.00        .00

    USLE SOIL FACTOR K FOR CHANNEL
 .310       .310      .310       .000

    USLE SOIL FACTOR C FOR CHANNEL
1.000      1.000     1.000       .000
```

SWRRB 1/01/89 IBM PC VERSION

CHICKASHA, OKLAHOMA
WATERSHED 522
LITTLE WASHITA RIVER BASIN

SOILS DATA

ST NO	LAYER DEPTH (MM)	POROSITY (MM/MM)	15 BAR SW (MM/MM)	.3 BAR SW (MM/MM)	AVAIL W ST (MM)	INITIAL W ST (MM)	SAT COND (MM/H)
1	10.0	.47	.09	.28	1.9	.5	16.0
2	150.0	.47	.09	.28	26.6	6.8	16.0
3	510.0	.42	.14	.33	68.4	17.6	15.0
4	840.0	.40	.16	.35	62.7	16.1	12.0
5	1190.0	.42	.22	.40	63.0	16.2	12.0
6	1520.0	.42	.19	.37	59.4	15.3	12.0
7	1900.0	.42	.19	.37	68.4	17.6	12.0
TOTALS					350.4	90.1	
1	10.0	.46	.09	.22	1.3	.3	50.0
2	150.0	.46	.11	.24	18.2	4.7	50.0
3	560.0	.46	.11	.19	32.8	8.4	50.0
4	760.0	.40	.12	.23	22.0	5.7	21.0
5	970.0	.40	.12	.25	27.3	7.0	21.0
6	1420.0	.40	.06	.15	40.5	10.4	21.0
7	1830.0	.40	.06	.14	32.8	8.4	21.0
TOTALS					174.9	45.0	
1	10.0	.46	.13	.28	1.5	.4	7.0
2	280.0	.46	.13	.28	40.5	10.4	5.0
3	330.0	.41	.21	.36	7.5	1.9	.8
4	530.0	.42	.23	.37	28.0	7.2	.8
5	760.0	.42	.27	.41	32.2	8.3	.8
6	1140.0	.42	.26	.40	53.2	13.7	.8
7	1905.0	.42	.26	.40	107.1	27.5	.8
TOTALS					270.0	69.4	
1	10.0	.46	.07	.26	1.9	.5	12.0
2	280.0	.46	.13	.32	51.3	13.2	12.0
3	560.0	.46	.22	.41	53.2	13.7	12.0
4	860.0	.45	.26	.44	54.0	13.9	12.0
5	1270.0	.45	.24	.42	73.8	19.0	12.0
6	1500.0	.45	.24	.42	41.4	10.6	12.0
7	1828.0	.45	.24	.42	59.0	15.2	12.0
TOTALS					334.6	86.0	

INITIAL COMPOSITE ST = 62.7 MM

SOIL SURFACE LAYER

SUB-BASIN	CLAY	SILT	SAND	K
1	.16	.65	.19	.31
2	.15	.35	.50	.25
3	.23	.58	.19	.35
4	.12	.72	.16	.31

SEDIMENT SIZE DISTRIBUTION

SUB-BASIN	SIZE(MM)					
	SAND	SILT	CLAY	SM AG	L AG	
	.20	.01	.002	.03	.50	
1	.123	.084	.032	.320	.440	
2	.334	.045	.030	.300	.291	
3	.099	.075	.046	.460	.319	
4	.116	.094	.024	.240	.526	1

```
SWRRB    1/01/89 IBM PC  VERSION

CHICKASHA, OKLAHOMA
WATERSHED 522
LITTLE WASHITA RIVER BASIN

CROP DATA


   SUB-BASIN
      1       PLANTING DATE =  3/15
              HARVEST DATE =  9/15
              TILLAGE OPERATION =FALL PLOW
              POT HEAT UNITS =  2286. C
              AVE C FACTOR =        .170
              MAX LAI =  4.0
              MAX ROOTING DEPTH =    1900.00 MM

   SUB-BASIN
      2       PLANTING DATE =  3/15
              HARVEST DATE =  9/15
              TILLAGE OPERATION =SPRG PLOW
              POT HEAT UNITS =  2286. C
              AVE C FACTOR =        .130
              MAX LAI =  4.0
              MAX ROOTING DEPTH =    1830.00 MM

   SUB-BASIN
      3       PLANTING DATE =  3/15
              HARVEST DATE =  9/15
              TILLAGE OPERATION =CONS TILL
              POT HEAT UNITS =  2286. C
              AVE C FACTOR =        .100
              MAX LAI =  4.0
              MAX ROOTING DEPTH =    1905.00 MM

   SUB-BASIN
      4       PLANTING DATE =  3/15
              HARVEST DATE =  9/15
              TILLAGE OPERATION =ZERO TILL
              POT HEAT UNITS =  2286. C
              AVE C FACTOR =        .050
              MAX LAI =  4.0
              MAX ROOTING DEPTH =    1828.00 MM
```

IRRIGATION DATA

SUBBASIN	IRRIGATE (1=YES,0=NO)	WATER STRESS	RUNOFF RATIO
1	0	.00	.00
2	0	.00	.00
3	0	.00	.00
4	0	.00	.00

PAGE 8

SWRRB 1/01/89 IBM PC VERSION

CHICKASHA, OKLAHOMA
WATERSHED 522
LITTLE WASHITA RIVER BASIN

	R (MM)	SURQ (MM)	SUB SUR Q (MM)	WATER YIELD (MM)	PERCO LATE (MM)	TRANS LOSSES (MM)	ET (MM)	Y (T/HA)	SW (MM)
1	2.09	.00	.00	.00	.29	.00	18.26	.00	46.20
2	15.49	.00	.00	.00	2.01	.00	7.58	.00	52.10
3	76.68	.48	.06	.33	7.05	.03	47.78	.03	73.38
4	69.59	.25	.05	.17	7.64	.02	36.74	.00	98.25
5	38.39	.38	.10	.29	4.00	.00	69.17	.01	63.04
6	49.30	.31	.05	.21	4.69	.01	97.45	.00	9.90
7	26.26	.00	.02	.02	2.74	.00	28.99	.00	4.43
8	20.38	.00	.01	.00	2.51	.00	20.54	.00	1.75
9	149.07	20.78	.24	14.52	12.18	.15	41.82	1.02	75.13
10	40.56	.00	.78	.70	5.32	.00	42.84	.00	67.00
11	18.04	.00	.39	.33	2.16	.00	21.05	.00	61.79
12	6.20	.00	.15	.11	1.37	.00	11.35	.00	55.26
1970	512.04	22.20	1.85	16.68	51.96	.21	443.56	1.08	55.26
1	12.43	.87	.07	.64	2.11	.01	8.24	.41	56.46
2	43.33	1.21	.04	.72	6.48	.00	20.94	.07	71.06
3	1.95	.00	.09	.05	.56	.00	10.07	.00	62.34
4	9.62	.00	.04	.02	2.56	.00	7.97	.00	61.41
5	122.79	9.14	.07	5.92	12.13	.07	56.38	.41	106.16
6	63.90	.09	.52	.38	6.76	.02	140.14	.00	22.62
7	59.83	.00	.25	.16	5.83	.00	54.15	.00	22.40
8	121.08	4.36	.43	3.18	11.36	.04	104.60	.06	22.51
9	135.76	5.30	.81	4.67	11.82	.18	48.10	.08	91.04
10	123.89	20.19	3.97	22.79	12.88	.06	57.12	1.15	119.84
11	16.94	.00	1.86	1.70	2.48	.00	19.80	.00	114.32
12	68.74	.65	1.80	2.08	9.30	.02	39.43	.02	131.34
1971	780.25	41.81	9.95	42.32	84.26	.42	566.94	2.20	131.34
1	1.68	.00	1.15	1.04	.73	.00	13.66	.00	118.56
2	12.22	.00	.42	.35	2.49	.00	10.20	.00	118.03
3	10.02	.00	.19	.12	1.36	.00	5.86	.00	120.80
4	136.48	8.44	1.00	6.93	16.89	.08	56.27	.33	169.41
5	85.54	12.45	5.14	16.39	10.06	.09	73.54	.38	155.61
6	25.66	.06	2.35	2.08	3.19	.00	170.68	.00	7.34
7	19.03	.00	.88	.71	2.33	.00	23.40	.00	.63
8	32.16	.00	.31	.23	4.19	.00	23.89	.00	4.72
9	30.23	.00	.11	.08	3.47	.00	24.18	.00	7.30
10	232.63	43.94	.53	36.35	17.82	.20	27.72	1.68	144.55
11	51.21	2.13	5.41	10.75	9.62	.02	46.82	.05	133.34
12	15.89	.00	2.71	2.44	2.49	.00	18.36	.00	128.21
1972	652.74	67.03	20.19	77.47	74.64	.40	494.57	2.45	128.21

PAGE 9

```
SWRRB    1/01/89 IBM PC  VERSION

CHICKASHA, OKLAHOMA
WATERSHED 522
LITTLE WASHITA RIVER BASIN
FINAL VALUES

SUB-BASIN                    SOIL WATER FOR LAYER NO
  NO
       1     2     3     4     5     6     7
  1   .0   8.3  68.4  56.0   9.0   4.3   2.7
  2   .0   1.2  13.2  22.0  27.3  40.5  10.0
  3   .0  24.9   7.6  28.3  32.2  28.8   1.4
  4   .0  36.4  53.2  54.0  16.2   5.9   3.4

    TOTAL SOIL WATER
  1      148.7
  2      114.3
  3      123.2
  4      169.0

       FINAL COMPOSITE ST =   128.2 MM
       MIN INDIVIDUAL WATER ST =        .0 MM


                     FINAL CONTENTS
          -----PONDS-----     ---RESERVOIRS---
          WATER     SED       WATER     SED
SUB-BASIN VOL       CONC       VOL      CONC
  NO       (MM)     (PPM)      (MM)     (PPM)
  1       11.0     400.1      23.4     250.0
  2       19.3     300.0      23.7     250.0
  3       12.5     450.4      27.3     250.4
  4       18.7     250.1        .0        .0
       FINAL COMPOSITE POND ST =   16.24 MM
       FINAL COMPOSITE RESERVOIR ST =   24.15 MM

                  IRRIGATION - AVE. ANNUAL
       SUB-BASIN         NO.OF              VOLUME
          NO.        APPLICATIONS        APPLIED(MM)

           1               0                 .000
           2               0                 .000
           3               0                 .000
           4               0                 .000

       SOIL WATER BALANCE =     -.127554E-03 MM

       POND BALANCE
       Q =      .366719E-05 MM     Y =      .131909E-05 T/HA

       RESERVOIR BALANCE
       Q =      .964357E-05 MM     Y =     -.835807E-06 T/HA
```

SWRRB 1/01/89 IBM PC VERSION

CHICKASHA, OKLAHOMA
WATERSHED 522
LITTLE WASHITA RIVER BASIN
SUB-BASIN STATISTICS

AVE ANNUAL VALUES

SUB-BASIN NO	RAIN (MM)	SUR Q (MM)	SUB SUR Q (MM)	Y (T/HA)	TOTAL BIOMASS (KG/HA)
1	634.2	13.5	7.3	2.9	14783.7
2	654.8	65.1	3.4	2.2	11224.3
3	656.8	22.4	77.4	.6	14705.1
4	646.0	19.4	8.8	.0	15361.4

AVE MONTHLY BASIN VALUES

MO	R (MM)	SNOW FALL (MM)	SUR Q (MM)	SUB SUR Q (MM)	WATER YIELD (MM)	ET (MM)	Y (T/HA)
1	5.40	.00	.29	.41	.56	13.38	.14
2	23.68	.00	.40	.16	.36	12.91	.02
3	29.55	.04	.16	.11	.17	21.24	.00
4	71.89	.00	2.90	.36	2.38	33.66	.11
5	82.24	.00	7.32	1.77	7.53	66.37	.27
6	46.29	.00	.16	.97	.89	136.09	.00
7	35.04	.00	.00	.38	.30	35.51	.00
8	57.87	.00	1.45	.25	1.14	49.68	.02
9	105.02	.00	8.69	.39	6.42	38.03	.37
10	132.36	.00	21.38	1.76	19.95	42.56	.94
11	28.73	.00	.71	2.55	4.26	29.22	.02
12	30.28	1.04	.22	1.55	1.54	23.04	.00

BASIN STATISTICS

```
        STD DEV OF RAIN STORMS
             13.897    14.571    15.738    15.486

    CN--MEAN = 66.597      MAX = 90.988      MIN = 56.290

    PRED PK FLOW
        MEAN =     49.831 M**3/S       ST DEV =     60.072 M**3/S
        MEAN PK/VOL =     16.261 M**3/S/MM      NO PKS =        25
        MAX =    216.792 M**3/S

    PRED MO WATER YLD
        MEAN =     3.79 MM
        ST DEV =     7.66 MM
```

PAGE 11

```
SWRRB    1/01/89 IBM PC  VERSION

CHICKASHA, OKLAHOMA
WATERSHED 522
LITTLE WASHITA RIVER BASIN
     AVE ANNUAL BASIN VALUES

          PRECIP =    648.3 MM
          SNOW FALL =    1.07 MM
          SNOW MELT =     .79 MM
          PRED SURFACE Q =    43.68 MM
          SUB-SUR Q =   10.67 MM
          PRED H20 YLD =    45.49 MM
          DEEP PERC =  70.29 MM
          ET =    501.7 MM
          TRANS LOSSES =       .34 MM
          TOTAL SUB-BASIN SED YLD =      2.156 T/HA
          BASIN SED YLD =    1.909 T/HA
          POND BUDGET
               EVAPORATION =  21.037 MM
               SEEPAGE =   16.577 MM
               RAINFALL ON POOL =  14.054 MM
               INFLOW
                    Q =   42.165 MM
                    Y =    3.189 T/HA
               OUTFLOW
                    Q =  19.857 MM
                    Y =     .378 T/HA
          RESERVOIR BUDGET
               EVAPORATION =   13.556 MM
               SEEPAGE =    9.830 MM
               RAINFALL ON POOL =    8.970 MM
               INFLOW
                    Q =  58.235 MM
                    Y =    2.695 T/HA
               OUTFLOW
                    Q =  42.313 MM
                    Y =     .292 T/HA
          YIELD LOSS FROM PONDS
               Q =  4.461 MM
               Y =    .535 T/HA
          YIELD LOSS FROM RESERVOIRS
               Q =  4.050 MM
               Y =    .577 T/HA
```

PAGE 12

CHICKASHA, OKLAHOMA
WATERSHED 522
LITTLE WASHITA RIVER BASIN

MONTHLY AND ANNUAL WATER STATISTICS
MONTHLY STATISTICS

	PRED	OBS	ERR
1	.00	1.35	-1.35
2	.00	1.42	-1.42
3	.33	2.95	-2.62
4	.17	3.81	-3.64
5	.29	2.08	-1.79
6	.21	.91	-.70
7	.02	.05	-.03
8	.00	.00	.00
9	14.52	2.84	11.68
10	.70	.64	.06
11	.33	.66	-.33
12	.11	.96	-.85
1970	16.68	17.67	-5.60
1	.64	1.19	-.55
2	.72	1.57	-.85
3	.05	1.11	-1.06
4	.02	.64	-.62
5	5.92	.91	5.01
6	.38	3.47	-3.09
7	.16	.25	-.09
8	3.18	2.54	.64
9	4.67	2.69	1.98
10	22.79	10.90	11.89
11	1.70	1.27	.43
12	2.08	2.89	-.81
1971	42.32	29.43	43.80
1	1.04	1.88	-.84
2	.35	1.57	-1.22
3	.12	1.57	-1.45
4	6.93	5.79	1.14
5	16.39	3.58	12.81
6	2.08	.71	1.37
7	.71	.22	.49
8	.23	.00	.23
9	.08	.05	.03
10	36.35	7.26	29.09
11	10.75	3.78	6.97
12	2.44	1.54	.90
1972	77.47	27.95	177.18

PAGE 13

MEAN MEAS =	.208472E+01	R**2 =	.594
MEAN PRED =	.379093E+01	REG SLOPE =	.223
STD DEV MEAS =	.221187E+01	SUM ER**2 =	.141590E+04
STD DEV PRED =	.766127E+01	NUM OBS =	36

	J	F	M	A	M	J	J	A	S	O	N	D
PRED	.56	.36	.17	2.38	7.53	.89	.30	1.14	6.42	19.95	4.26	1.54
OBS	1.47	1.52	1.88	3.41	2.19	1.70	.17	.85	1.86	6.27	1.90	1.80

ANNUAL STATISTICS

	PRED	OBS	ERR
1970	16.68	17.67	-.99
1971	42.32	29.43	12.89
1972	77.47	27.95	49.52

MEAN MEAS =	.250167E+02	R**2 =	.556
MEAN PRED =	.454912E+02	REG SLOPE =	.156
STD DEV MEAS =	.640529E+01	SUM ER**2 =	.261954E+04
STD DEV PRED =	.305190E+02	NUM OBS =	3

CHICKASHA, OKLAHOMA
WATERSHED 522
LITTLE WASHITA RIVER BASIN

MONTHLY AND ANNUAL SEDIMENT STATISTICS
MONTHLY STATISTICS

	PRED	OBS	ERR
1	.00	.02	-.02
2	.00	.02	-.02
3	.03	.05	-.02
4	.00	.24	-.23
5	.01	.04	-.03
6	.00	.00	-.00
7	.00	.00	.00
8	.00	.00	.00
9	1.02	.26	.76
10	.00	.00	-.00
11	.00	.00	-.00
12	.00	.00	-.00
1970	1.08	.67	60.98
1	.41	.00	.40
2	.07	.02	.05
3	.00	.00	-.00
4	.00	.00	-.00
5	.41	.07	.34
6	.00	.41	-.41
7	.00	.00	.00
8	.06	.20	-.14
9	.08	.22	-.14
10	1.15	1.33	-.18
11	.00	.00	-.00
12	.02	.08	-.06
1971	2.20	2.37	-7.07
1	.00	.02	-.02
2	.00	.02	-.02
3	.00	.02	-.02
4	.33	.76	-.43
5	.38	.11	.27
6	.00	.00	-.00
7	.00	.00	.00
8	.00	.00	.00
9	.00	.00	.00
10	1.68	.90	.78
11	.05	.14	-.09
12	.00	.02	-.02
1972	2.45	2.00	22.28

MEAN MEAS	= .140000E+00	R**2	=	.584
MEAN PRED	= .159075E+00	REG SLOPE	=	.588
STD DEV MEAS	= .287024E+00	SUM ER**2	=	.204114E+01
STD DEV PRED	= .373333E+00	NUM OBS	=	36

	J	F	M	A	M	J	J	A	S	O	N	D
PRED	.14	.02	.00	.11	.27	.00	.00	.02	.37	.94	.02	.00
OBS	.02	.02	.03	.34	.07	.14	.00	.07	.16	.75	.05	.04

ANNUAL STATISTICS

	PRED	OBS	ERR
1970	1.08	.67	.41
1971	2.20	2.37	-.17
1972	2.45	2.00	.45

MEAN MEAS	= .168000E+01	R**2	=	.865
MEAN PRED	= .190889E+01	REG SLOPE	=	1.140
STD DEV MEAS	= .894036E+00	SUM ER**2	=	.393548E+00
STD DEV PRED	= .729273E+00	NUM OBS	=	3

Sensitivity Analysis

Sensitivity analysis is used to assess the relative change in model output resulting from a change in inputs or model parameters. For complex simulation models, sensitivity is often expressed in the form of relative changes, differentials, and graphs.

The approach taken here is similar to the sensitivity analysis performed on CREAMS (Chapter 6. Sensitivity Analysis, L. J. Lane and V. A. Ferreira in Knisel, 1980). Base values are selected based on representative values of the input variables. Computations are performed with the selected set of base input variables. With all other input variables at their base values, each individual input variable is allowed to deviate from its base and the computations are repeated. The resulting computations show how model outputs vary with changes in the input variables.

Complex interactions are difficult to determine since the parameters are varied individually. Also, the model is fairly comprehensive and a change in one component of the hydrologic budget will effect other components. For example, an increase in surface runoff will cause an increase in peak rates and consequently an increase in sediment yields. It will also cause other parts of the hydrologic balance, such as ET, return flow or percolate to decrease depending on the specific conditions. Biomass and crop yields may also decrease. The effect on basin sediment yields will be hard to determine because of the effects of sediment routing. Depending on subsurface conditions, return flow may pick up a large share of the increased infiltration, causing little change in basin water yield. Pond and reservoir storage will also affect surface runoff and consequently basin water yield. Obviously, the interactions are complex and comprehensive, and care must be taken when interpreting output results.

The sensitivity analysis is limited to a subset of the input variables because the number of simulations runs increase rapidly with the number of input variables and the number of points selected to vary about the base values. Only the major components and the input variables they are most sensitive to are analyzed. Each simulation run consisted of a five-year period to obtain a representative sample. The sensitivity analyses presented here are for particular locations and may vary greatly from location to location. This analysis is intended as an introduction and users should determine model sensitivity for their particular application.

Surface Runoff and Peak Rates

Surface runoff (Q) is a direct function of soil water. All hydrologic components either directly or indirectly affect soil water and thus affect Q. Two locations were chosen to test the sensitivity of Q and peak rates (q_p) to select model inputs. Condition II CN and soil data (BD, AWC, and SC) were chosen at Riesel, Texas, while LAI_{mx}, temperature, and solar radiation were chosen at Coshocton, Ohio (Table 1). Since Q is calculated with a curve number approach, it is most sensitive to CN_2. BD and AWC have a large effect on soil water storage and consequently have a relatively large effect on Q as shown in Table 1 at Riesel. Q is not as sensitive to SC in this case, which is fortunate since SC has the largest amount of variability and uncertainty. Since Q at Riesel is high, it may be more sensitive to these input variables than other locations. Q at Coshocton is relatively insensitive to LAI_{mx}, temperature, and solar radiation. These inputs have more influence on ET and biomass and in this particular situation have little effect on Q and q_p.

Sensitivity of q_p to the input variables is less than Q but generally follows the same trends. Since q_p has a stochastic component, there may be small unexplained movements.

Subsurface Flow

Subsurface flow (QR) is most sensitive to an impermeable soil layer thickness, and the baseflow factor. It is also closely dependent on changes in other components of the hydrologic balance. QR is most sensitive to an impermeable layer at the bottom of the soil profile. Two five-year simulation runs were performed at Coshocton with and without an impermeable soil layer, resulting in QR rates of 260 and 92 mm/year, respectively. QR is also sensitive to the baseflow factor (BFF) as shown in Table 2. Base values of BFF at Coshocton and Riesel are 1.0 and 0.10, respectively, and the model is not very sensitive around the base values. Finally, QR is slightly sensitive to layer thickness. Table 3 for Coshocton shows that as layer thickness increases from 100 to 900 mm, QR decreases by 10%.

ET and Biomass

Coshocton and Riesel were selected to test the sensitivity of ET and biomass to selected model inputs (Table 4). LAI_{mx}, temperature, solar radiation, and root depth were chosen at Coshocton while CN_2 and AWC were chosen at Riesel. The inputs selected at Coshocton did not have extreme influence on ET and only slightly greater influence on biomass.

ET and biomass were more sensitive to CN_2 and AWC at Riesel. CN_2 affects ET and biomass indirectly by affecting Q, thus altering the entire hydrologic balance. AWC also impacts ET and biomass by limiting plant available water. Increasing CN_2 and AWC could have a double impact on sediment yields. Increasing CN_2 and decreasing AWC would increase Q and consequently sediment yields. It would also decrease biomass, thus increasing the C factor and increasing sediment yields.

Transmission Losses

The sensitivity of transmission losses (TL) to selected inputs was tested at Tombstone, Arizona, where TL are significant in the hydrologic balance (Table 5). TL are most sensitive to CHK, but are also sensitive to stream channel inputs such as width, depth, and length. TL may also be sensitive to the number of basin subdivisions.

Ponds and Reservoirs

A reservoir study by Williams et al. (1985) shows the model's sensitivity of basin water and sediment yields to reservoir storage. The study also demonstrates the model's usefulness in evaluating the benefits of the structures.

During 1969-1971, 19 SCS flood control structures were constructed to control runoff from 17% of the Little Washita basin near Chickasha, Oklahoma (Watershed OK-522). Simulations were performed for each of three basin conditions: 1) before flood control, 2) after structures were installed, and 3) assuming no structures for the period after structures were built.

Table 6 shows measured and simulated water and sediment yields for periods with and without structures. Note the close agreement between means and standard deviations for both runoff and sediment yield. It is also interesting to compare the information contained in Table 6 to estimate the effects of structures on water and sediment yield. However, it is almost impossible to estimate structural effects using Table 6 because of the large difference in rainfall. Structural effectiveness can be estimated, however, by simulating the period after structures were installed assuming no structures (the third condition simulated). If structures had not been in place, the average annual water and sediment yield during 1971-1981 would have been 71 mm and 7.2 t/ha according to the SWRRB simulation. By comparing these simulation results with those shown in Table 6, the structures can be evaluated. From this comparison, the structures appear to be quite beneficial--water yield was reduced by only 4 mm/year and sediment yield was reduced by 1.6 t/ha/year. Thus, the structures did not significantly deplete water yield to downstream reservoirs, but they did trap about 28% of the sediment that would have been deposited downstream (Williams et al., 1985).

Water Yield

It should be obvious by now that all model components are interrelated and these interactions are sensitive to the input variables. Basin water yield is the sum of Q and QR and is also directly affected by transmission losses and pond and reservoir storage. Any input variables that directly affect these components will probably affect basin water yields. Also, a change in other components of the hydrologic budget may also affect basin water yield. Because of the variability and complexity involved with basin water yield, users should determine model sensitivity for their particular application.

Sediment Yields

Sensitivity of sediment yields to selected inputs was examined at Chickasha, Oklahoma (Table 7). Since subbasin sediment yield is computed as a function of subbasin Q and q_p, it is sensitive to changes in all of the inputs affecting Q and q_p. Subbasin sediment yields are also sensitive to the MUSLE parameters C, PE, K, and LS. There is a linear relationship between PE, K, and LS versus subbasin sediment yields and these inputs remain constant throughout the year. However, C is allowed to vary thoughout the year as a function of residue and live biomass cover. Sediment yield at Chickasha is very sensitive to variation in C as shown in Table 7. Sediment yield is also sensitive to the tillage operation, or amount of residue left on the surface, since it has a direct effect on C.

When multiple subbasins are simulated, basin sediment yield is also affected by channel input variables which influence sediment routing. The effect of w, S_c, L_c, and n on basin sediment yield is shown in Table 7 and in some cases can be significant.

Table 1. Sensitivity of surface runoff and peak rates to selected inputs.

Run number	Variation	Parameter	Mean runoff volume Q	Q/Base	Mean peak discharge q_p	q_p/base
	(%)		(mm)		(m^3/s)	
Riesel						
1	Base		107	1.00	5.86	1.00
2	-10		38	0.36	4.66	0.80
3	-5	CN_2	64	0.60	5.31	0.91
4	+5		176	1.64	8.76	1.49
5	+10		308	2.88	11.73	2.00
6	-50		149	1.39	7.83	1.34
7	-25	BD	137	1.28	7.82	1.33
8	+25		52	0.49	5.68	0.97
9	+50		51	0.48	5.57	0.95
10	-50		167	1.56	8.10	1.38
11	-25	AWC	136	1.27	7.83	1.34
12	+25		78	0 .73	5.87	1.00
13	+50		54	0.50	5.71	0.97
14	-50		106	0.99	5.95	1.02
15	-25	SC	107	1.00	5.86	1.00
16	+25		107	1.00	5.87	1.00
17	+50		107	1.00	5.88	1.00
Coshocton						
18	Base		81.2	1.00	5.97	1.00
19	-50		84.6	1.04	5.83	0.98
20	-25	LAI_{mx}	83.3	1.03	5.60	0.94
21	+25		80.4	0.99	5.82	0.97
22	+50		78.3	0.96	5.87	0.98
23	-25		81.4	1.00	6.27	1.05
24	-12.5	TEMP	82.8	1.01	6.46	1.08
25	+12.5		81.3	1.00	5.23	0.88
26	+25		80.1	0.99	5.15	0.86
27	-20	RA	91.3	1.12	6.16	1.03
28	-10		86.9	1.07	5.65	0.95
29	+10		77.4	0.95	5.78	0.97
30	+20		72.5	0.89	5.51	0.92

Table 2. Sensitivity of return flow to the baseflow factor.

Run number	BFF	Coshocton QR	Riesel QR
		(mm)	(mm)
1	1.00	260	27.7
2	0.75	256	22.6
3	0.50	247	16.8
4	0.25	234	9.8
5	0.10	202	4.4
6	0.00	77	0.5

Table 3. Sensitivity of return flow to layer thickness at Coshocton.

Run number	Number of layers	Thickness	QR
		(mm)	(mm)
1	2	900	291
2	3	450	294
3	5	225	307
4	10	100	323

Table 4. Sensitivity of ET and biomass to selected inputs.

Run number	Variation	Parameter	ET	ET/Base	Biomass	DM/Base
	(%)		(mm)		(t/ha)	
Coshocton						
1	Base		635	1.00	8.20	1.00
2	-50		623	0.98	5.65	0.69
3	-25	LAI_{mx}	631	0.99	6.89	0.84
4	+25		639	1.01	9.19	1.12
5	+50		641	1.01	9.79	1.19
6	-25		616	0.97	7.46	0.91
7	-12.5		633	1.00	8.91	1.09
8	+12.5	TEMP	645	1.02	7.84	0.96
9	+25		658	1.04	7.53	0.92
10	-20		592	0.93	7.07	0.86
11	-10	RA	618	0.97	7.72	0.94
12	+10		654	1.03	8.90	1.09
13	+20		666	1.05	9.35	1.14
14	-50	Maximum	605	0.95	6.81	0.83
15	-25	root depth	630	0.99	8.13	0.99
Riesel						
16	Base		573	1.00	10.46	1.00
17	-10		626	1.09	12.25	1.17
18	-5	CN_2	606	1.06	11.57	1.11
19	+5		513	0.90	8.25	0.79
20	+10		391	0.68	3.54	0.34
21	-50		631	1.10	12.41	1.19
22	-25	AWC	604	1.05	11.50	1.10
23	+25		541	0.94	9.34	0.89
24	+50		508	0.89	7.86	0.75

Table 5. Sensitivity of transmission losses to selected inputs.

Run number	Variation	Parameter	TL	TL/Base	Water yield	Water yield/ base
	(%)		(mm)		(mm)	
Tombstone						
1	Base		3.97	1.00	5.05	1.00
2	-50		2.53	0.64	6.33	1.25
3	-25	CHK	3.36	0.85	5.58	1.11
4	+25		4.60	1.15	4.49	0.89
5	+50		5.18	1.30	4.00	0.79
6	-50		3.34	0.84	5.60	1.11
7	-25	w	3.67	0.92	5.31	1.05
8	+25		4.24	1.07	4.80	0.95
9	+50		4.50	1.13	4.58	0.91
10	-25		3.67	0.92	5.31	1.05
11	-12.5	L_c	3.82	0.96	5.17	1.02
12	+12.5		4.11	1.04	4.92	0.97
13	+25		4.24	1.07	4.80	0.95

Table 6. Comparison of measured and SWRRB simulated water and sediment yields with and without structures.*

	Rainfall	Runoff		Sediment yield	
		Measured	Simulated	Measured	Simulated
	(mm)			(t/ha)	
Without structures (1962-1970)					
Mean	649	35	42	2.8	2.7
Standard deviation	142	19	23	2.4	1.6
With structures (1971-1981)					
Mean	779	60	67	4.5	4.6
Standard deviation	163	29	35	2.6	2.5

* Taken from Williams et al. (1985).

Table 7. Sensitivity of sediment yields to selected inputs.

Run number	Variation	Parameter	Y	Y/Base
	(%)		(t/ha)	
Chickasha				
1	Base		4.4	1.00
2	-50		1.7	0.38
3	-25	C	2.9	0.66
4	+25		6.4	1.45
5	+50		8.6	1.95
6	Fall plow	Tillage	12.5	2.84
7	Cons. till	operation	2.4	0.55
8	Zero till		1.9	0.43
9	Base		2.9	1.00
10	-50		5.0	1.72
11	-25	w	3.5	1.21
12	+25		2.5	0.86
13	+50		2.3	0.79
14	-50		2.5	0.86
15	-25	S_c	2.7	0.93
16	+25		3.0	1.03
17	+50		3.2	1.10
18	-50		3.2	1.10
19	-25	L_c	3.0	1.03
20	+25		2.8	0.97
21	+50		2.7	0.93
22	-50		2.6	0.90
23	-25	n	2.7	0.93
24	+25		3.0	1.03
25	+50		3.1	1.07

Model Evaluation

SWRRB has been tested on 11 large watersheds from 8 ARS locations throughout the U.S. (Arnold and Williams, 1987). Table 8 gives a brief description of the watersheds. Table 9 shows the results of measured and simulated annual water yields. It is important for simulation models to produce frequency distributions that are similar to measured frequency distributions. Close agreement between means and standard deviations indicates that the frequency distributions are similar. Generally, simulated values compare very well with measured values.

Comparing monthly water and sediment yields allows us to examine the accuracy of seasonal phenomena such as climate, crop growth, residue cover, etc. The results of measured and simulated monthly water yields are shown in Table 10. Regression line slopes and R^2 values near unity indicate a close relationship between measured and simulated monthly water yields. Monthly standard deviations are given in Table 10 to indicate the agreement in monthly frequency distributions. In general, the simulated monthly water yields compare well with measured monthly values.

The annual and monthly results of measured and simulated sediment yields are shown in Table 11. Only five of the selected watersheds had sediment data available. Means, standard deviations, and R^2 values are given to indicate agreement in frequency distributions. The simulated sediment yields compare reasonably well with measured sediment yields considering the wide range of conditions.

Peak runoff rates are based on a modification of the Rational Formula (Williams et al., 1985). Since SWRRB is primarily a long-term water and sediment yield simulator, a high degree of accuracy in predicting peak rates is not as necessary as far as for other applications like flood control planning and flood forecasting. However, peak rates are used in MUSLE (Williams and Berndt, 1977) for computing sediment yields. Maximum annual peak rates were obtained for watersheds OK-522, TX-G, ID-1, and MS-34 and compared with SWRRB simulated peak rates. Results are shown in Table 12. Means and standard deviations compare well.

Savabi et al. (1988) and Krishna et al. (1988) tested the SWRRB water balance for crop and range conditions on several small watersheds at Riesel, Texas, and found good agreement. The crop model was tested against measured crop yields by Arnold and Stockle (1988). SWRRB was able to estimate water use efficiency for various irrigation treatments and was also able to predict the effect of management decisions on water and sediment yields (Arnold and Williams, 1985). Williams et al. (1985) utilized SWRRB to demonstrate the effects of reservoir storage on water and sediment yields on a large watershed at Chickasha, Oklahoma. The simulation results showed the structures to be quite beneficial--water yield was reduced by only 4 mm per year and sediment yield was reduced by 1.6 t/ha per year.

SWRRB was utilized to determine the effect of urbanization on water and sediment entering White Rock Lake in Dallas (Arnold et al., 1987). The simulation results show that, if urbanization had not occurred, annual surface runoff would be 135 mm rather than 151 mm and annual sediment yield would be 4.4 t/ha rather than 4.1 t/ha. The effect of urbanization on delivery ratios was shown and a positive linear correlation was found. Finally, the weather generator in SWRRB was utilized to estimate the loss of reservoir capacity until 2050 for three different land use management scenarios.

Table 8. Watersheds for SWRRB validation.

Location	ARS watershed	Area	Ave. land slope	No. of subbasins	Average land use
		(km^2)	(m/m)		
Chickasha, OK	552	538.1	0.08	4	66% pasture & range, 18% crop, 16% misc.
Riesel, TX	G	17.7	0.02	1	65% crop, 22% pasture, 13% grass
Coshocton, OH	994	70.8	0.05	1	55% grass, 28% forest, 17% crop
Coshocton, OH	97	18.5	0.05	1	50% grass, 30% forest, 20% crop
Tifton, GA	B	334.0	0.02	2	42% forest, 38% crop, 20% pasture
Tifton, GA	K	16.7	0.02	2	56% forest, 30% crop, 14% pasture
Boise, ID	1	233.6	0.09	3	95% rangeland, 2% forest, 3% crop
Oxford, MS	34	303.5	0.05	5	55% pasture, 23% crop, 22% forest
N. Danville, VT	5	111.2	0.10	3	67% forest, 20% crop, 13% grass

Table 9. Comparison of measured and SWRRB simulated annual water yields.

ARS watershed	Period of record	Measured mean	Predicted mean	Measured st. dev.	Predicted st. dev.	Average % error
		(mm)				
OK-522	1962-81	48	49	27	25	31
TX-G	1961-82	169	149	129	88	30
OH-994	1942-65	334	338	107	110	13
OH-97	1944-53	296	284	95	94	14
GA-B	1972-80	352	352	58	58	11
GA-K	1972-80	416	402	68	54	8
ID-1	1967-80	75	76	40	37	24
MS-34	1964-76	445	446	161	150	16
VT-5	1960-73	556	521	133	146	11

Table 10. Comparison of measured and SWRRB simulated monthly water yields.

ARS watershed	Measured st. dev.	Predicted st. dev.	R^2	Regression slope
	(mm)			
OK-522	4.9	5.5	0.60	0.70
TX-G	27	20	0.78	1.14
OH-994	33	31	0.72	0.89
OH-97	30	21	0.55	1.06
GA-B	38	31	0.78	1.10
GA-K	41	32	0.63	1.02
ID-1	8.4	8.5	0.65	0.80
MS-34	38	42	0.83	0.82
VT-5	46	54	0.35	0.51

Table 11. Comparison of measured and SWRRB simulated sediment yields.

ARS watershed	Period of record	Annual Meas. mean	Annual Pred. mean	Annual Meas. st. dev.	Annual Pred. st. dev.	Monthly Meas. st. dev.	Monthly Pred. st. dev.	R^2
		(t/ha)						
OK-522	1964-81	3.8	3.4	2.6	2.7	.75	.85	.78
TX-G	1961-82	1.5	1.8	1.4	1.1	.32	.30	.75
GA-K	1979-80	.05	.08	.005	.01	.006	.01	.51
ID-1	1975-80	.26	.23	.15	.18	.05	.06	.17
MS-34	1964-75	13.6	6.6	9.0	2.4	2.0	.75	.58

Table 12. Comparison of measured and SWRRB simulated maximum annual peak rates.

Watershed	Years of record	Mean	Standard deviation	Maximum	Minimum
		(m^3/s)			
OK-522	12				
Measured		103	58.5	213	22.8
Predicted		114	72.7	261	21.1
TX-G	17				
Measured		37.8	31.3	118.6	0.5
Predicted		39.0	23.8	92.5	8.0
ID-1	9				
Measured		13.2	7.3	26.4	4.9
Predicted		18.9	12.8	33.9	6.0
MS-34	10				
Measured		297	84	356	156
Predicted		303	128	508	123

References

Arnold, J. G. and C. O. Stockle. 1988. Simulation of supplement irrigation from on-farm ponds. Draft.

Arnold, J. G. and J. R. Williams. 1985. Evapotranspiration in a basin scale hydrologic model. Proc. Natl. Symp. on Advances in Evapotranspiration. ASAE, December 16-17, 1985, Chicago, IL.

Arnold, J. G. and J. R. Williams. 1987. Validation of SWRRB--Simulator for water resources in rural basins. J. Water Resources Planning and Manage., ASCE, 113(2):243-256.

Arnold, J. G., M. D. Birckett, J. R. Williams, W. F. Smith, and H. N. McGill. 1987. Modeling the effects of urbanization on basin water yield and reservoir sedimentation. Water Resources Bull. 23(6):1101-1107.

Bagnold, R. A. 1977. Bedload transport by natural rivers. Water Resources Res. 13(2):303-312.

Boughton, W. C. 1984. A simple model for estimating the water yield on unguaged catchments. Civil Eng. Trans. Inst. of Eng., Australia. Paper No. C1454.

Chow, V. T. 1959. Open-Channel Hydraulics. McGraw-Hill Book Co., Inc. pp. 112-113.

Doorenbos, J. and A. H. Kassam. 1979. Yield response to water. Irrigation and Drainage Paper 33. Food and Agric. Org. of United Nations, Rome.

Engman, E. T. 1983. Roughness coefficients for routing surface runoff. Proc. ASCE Speciality Conf. Frontiers of Hydraulic Eng.

Foster, G. R., L. J. Lane, J. D. Nowlin, J. M. Laflen, and R. A. Young. 1980. A model to estimate the sediment yield from filed-sized areas: Development of model. CP-80-10. Intl. Inst. for Applied Systems Analysis. A-2361. Laxenburg, Austria.

Freeze, R. A. 1972. Rate of subsurface flow on generating surface runoff, 2, Upstream source areas. Water Resources Res. 8(5):1272-1283.

Freeze, R. A. and J. A. Cherry. 1979. Groundwater. Prentice-Hall, Inc., Englewood Cliffs, NJ. pp. 32-34.

Hershfield, D. M. 1961. Rainfall frequency atlas of the United States for durations from 30 minutes to 24 hours and return periods from 1 to 100 years. U.S. Dept. Commerce Tech. Paper No. 40. 115 pp.

Jones, C. A. 1985. C-4 Grasses and Cereals. John Wiley & Sons, Inc., New York. 419 pp.

Knisel, W. G. (ed.). 1980. CREAMS, A field scale model for chemicals, runoff, and erosion from agricultural management systems. USDA Conserv. Res. Rept. No. 26. 643 pp.

Krishna, J. H., J. G. Arnold, and C. W. Richardson. 1988. Predicting water and sediment yields from agricultural and grassland watersheds. ASAE Intl. Symp. Modeling Agricultural, Forest, and Rangeland Hydrology, Chicago, IL, December 1988.

Lane, L. J. 1982. Distributed model for small semi-arid watersheds. J. Hydraulic Eng., ASCE, 108(HY10):1114-1131.

Monsi, M. and T. Saeki. 1953. Uber den Lictfaktor in den Pflanzengesellschaften und sein Bedeutung fur die Stoffproduktion. Japan J. Bot. 14:22-52.

Monteith, J. L. 1977. Climate and the efficiency of crop production in Britain. Phil. Trans. Res. Soc. London B. 281:277-329.

Nicks, A. D. 1974. Stochastic generation of the occurrence, pattern, and location of maximum amount of daily rainfall. pp. 154-171 In Proc. Symp. on Statistical Hydrology, Aug.-Sept. 1971, Tucson, AZ. Misc. Publ. No. 1275.

Nicks, A. D. and J. F. Harp. 1980. Stochastic generation of temperature and solar radiation data. J. Hydrology 48:1-17.

Nicks, A. D. and F. A. Igo. 1980. A depth-area-duration model of storm rainfall in the Southern Plains. Water Resources Res. 16(5):939-945.

Priestley, C. H. B. and R. J. Taylor. 1972. On the assessment of surface heat flux and evaporation using large-scale parameters. Mon. Weather Rev. 100:81-92.

Richardson, C. W. and J. T. Ritchie. 1973. Soil water balance for small watersheds. Trans. ASAE 16(1):72-77.

Ritchie, J. T. 1972. A model for predicting evaporation from a row crop with incomplete cover. Water Resources Res. 8(5):12044-1213.

Savabi, M. R., J. G. Arnold, and C. W. Richardson. 1988. Modeling the effect of brush control on rangeland water yield. ASAE Intl. Symp. Modeling Agricultural, Forest, and Rangeland Hydrology, Chicago, IL, December 1988.

Sloan, P. G. and I. D. Moore. 1984. Modeling subsurface streamflow on steeply sloping forested watersheds. Water Resources Res. 20(12):1815-1822.

Smith, R. E. and R. H. B. Hebbert. 1983. Mathematical simulation of interdependent surface and subsurface hydrologic processes. Water Resources Res. 19(4):987-1001.

Svetlosanov, V. and W. G. Knisel. 1982. European and United States Case Studies in Application of the CREAMS Model. Intl. Inst. for Applied Systems Analysis. Bo. No. 3-7045-0056-9. Laxenburg, Austria. p. 126.

USDA, Soil Conservation Service. 1957. Farm Pond Survey-Texas. 12 pp.

USDA, Soil Conservation Service. 1972. National Engineering Handbook. Hydrology Section 4, Chapters 4-10, 19.

USDA, Soil Conservation Service. 1975. Procedures for computing sheet and rill erosion on project areas. Tech. Release No. 51, Geology 16.

U.S. Department of Commerce. 1978. Climatological data. National Summary. Annual Summary 29(13):98 pp.

U.S. Department of Commerce. 1979. Maximum short duration rainfall. National Summary, Climatic Data.

Water Information Center, Inc. 1974. Climate of the States, Vol. 1 Easter States, Vol. 2 Western States. Port Washington, NY. 479 pp.

Williams, J. R. 1980. SPNM, a model for predicting sediment, phosphorus, and nitrogen yields from agricultural basins. Water Resources Bull. 16(5):843-848.

Williams, J. R. 1986. EPIC, the erosion-productivity impact calculator. Volume I, Model documentation. Submitted as USDA-ARS Series.

Williams, J. R. and H. D. Berndt. 1976. Determining the universal soil loss equation's length-slope factor for watersheds. pp. 217-225 In Soil Erosion: Prediction and Control. Soil Conserv. Soc. Amer.

Williams, J. R. and H. D. Berndt. 1977. Sediment yield prediction based on watershed hydrology. Trans. ASAE 20(6):1100-1104.

Williams, J. R. and R. W. Haan. 1978. Optimal operation of large agricultural watersheds with water quality constraints. Texas Water Resources Inst., Texas A&M Univ., TR-96. 152 pp.

Williams, J. R. and W. V. Laseur. 1976. Water yield model using SCS curve numbers. J. Hydraulic Eng., ASCE, 102(HY9):1241-1253.

Williams, J. R. and A. D. Nicks. 1982. CREAMS hydrology model--Option one. pp. 69-86 In V. P. Singh (ed.) Applied Modeling Catchment Hydrology. Proc. Intl. Symp. Rainfall-Runoff Modeling, May 18-21, 1981, Mississippi State, MS.

Williams, J. R., C. A. Jones, and P. T. Dyke. 1984. A modeling approach to determining the relationship between erosion and soil productivity. Trans. ASAE 27(1):129-144.

Williams, J. R., A. D. Nicks, and J. G. Arnold. 1985. Simulator for water resources in rural basins. J. Hydraulic Eng., ASCE, 111(6):970-986.

Wischmeier, W. H. and D. D. Smith. 1978. Predicting rainfall erosion losses--A guide to conservation planning. USDA Agric. Handbook No. 537. 58 pp.

Appendix I

Symbol Definition

Symbol Definition

α = maximum amount of rainfall during t_c divided by the total rainfall for the day

α_{mn} = minimum value of α

α_{sp} = parameter dependent on maximum stream power

$\alpha_{.5}$ = maximum amount of rainfall during 0.5 h divided by the total rainfall for the day

$\alpha_{.5P}$ = $\alpha_{.5}$ value at the peak of the gamma distribution

$\alpha_{.5u}$ = upper limit of $\alpha_{.5}$

β_i = parameter for regulating hydraulic conductivity in soil layer i

β_r = routing coefficient

γ = psychrometric constant

γ_w = density of water (t/m^3)

δ = temperature stress parameter

δ_s = slope of saturation vapor pressure curve

ζ_i = soil parameter expressing degree of cracking for layer i

η = pond evaporation coefficient (≈ 0.6)

θ = sun's declination angle (radians)

λ = land surface slope length (m)

Λ = water use rate-depth parameter (3.065)

μ_1, μ_2 = lateral flow travel time-soil water parameters

ν = gamma distribution shape parameter for estimating $\alpha_{.5}$

ξ = MUSLE LS factor parameter (function of slope)

ρ = rational formula runoff coefficient

σ = standard deviation of daily precipitation (mm)

τ = total number of maximum 0.5 h rainfall events considered

ϕ = sun's half day length (radians)

χ = uniform random number (0-1)

$\omega 1$, $\omega 2$ = parameters in pond and reservoir surface area--volume relationship

Ω = average channel slope (m/m)

a_r = unit channel regression intercept for transmission losses (m^3)

a_x = regression intercept for transmission losses (m/km)

A = drainage area (ha)

AB = albedo

AB_s = soil albedo

AIR = volume of irrigation water applied (mm)

AM = annual amplitude in daily average temperature (°C)

AO = circular area used in precipitation adjustment for subbasins (km^2)

b = TP-40 rainfall distribution parameter

bcv = lagging factor for simulating residue and snow cover effect on soil surface temperature

b_r = unit channel regression slope for transmission losses

b_x = regression slope for transmission losses

B = accumulated plant biomass (above ground and roots) (kg/ha)

B_{AG} = above ground biomass (kg/ha)

B_F = weighting factor based on wet-dry rainfall probabilities

B_P = potential crop biomass (kg/ha)

B_t = fraction on three-part hydrograph in the first segment

B_{TOT} = total biomass (kg/ha)

BD = soil bulk density (t/m^3)

BE = crop parameter--converts energy to biomass (kg m^2/ha MJ)

BFF = baseflow factor (0-1)

BLT = basin lag time (days)

BRT = storage coefficient for basin lag time

BS = basin storage (mm)

c_I = inflow sediment concentration (t/m^3)

c_o = sediment concentration of outflow from ponds and reservoirs (t/m^3)

c_s = sediment concentration in ponds and reservoirs (t/m^3)

c_{Se} = equilibrium P/R sediment concentration (t/m^3)

C = MUSLE crop management factor

CDG = residue decay factor for soil water content

CHK = effective hydraulic conductivity of channel alluvium (mm/hr)

CI = contour interval (m)

CLA = fraction of primary clay in the original soil mass (0-1)

CLA_i = clay content of soil layer i (fraction)

CN = SCS runoff curve number

CN_1 = SCS curve number for the I moisture (dry) condition

CN_2 = CN for the II moisture (average) condition

CN_3 = CN for the III moisture (wet) condition

CV = soil cover (above ground biomass + residue) (kg/ha)

CVA = average annual value of the MUSLE crop management factor

CVM = minimum value of the MUSLE crop management factor

d_{bi} = distance from the bottom of the soil profile to the top

d_i = thickness of soil layer i (mm)

d_j = sediment particle diameter for size j (mm)

d_w = channel flow depth (m)

dm_j = number of days in month j

dp_j = number of days of precipitation in month j

D = earth's radius vector

D50 = median particle size of the inflow sediment (μm)

DA = drainage area (km^2)

DD	= soil temperature damping depth (mm)
DECR	= daily decrease in surface residue (0-1)
DEG	= total degradation (t)
DEG_B	= degradation of the bed material (t)
DEG_R	= reentrained sediment (t)
DEP	= sediment deposition (t)
DLAI	= fraction of growing season when LAI starts declining
DP	= maximum damping depth for soil temperature (mm)
DR	= sediment delivery ratio from subbasin to basin outlet
DR_{SB}	= sediment delivery ratio within the subbasin
DU	= duration of streamflow (h)
DUR	= rainfall duration (h)
E_o	= potential evaporation rate (mm/d)
E_P	= potential plant evaporation rate (mm/d)
E_s	= soil evaporation rate (mm/d)
E_{so}	= potential evaporation rate at the soil surface (mm/d)
EA	= soil cover index (0-1)
EFI	= irrigation runoff ratio
EP	= number of contour extreme points in the subbasin
ET	= evapotranspiration (mm)
EV	= evaporation from farm ponds (m^3)
f	= average infiltration rate (mm/h)
F	= frequence of occurrence of the largest 0.5 h rainfall recorded
FC_i	= field capacity (0.3 bar-15 bar) water content of soil layer i (mm)
FHU	= harvest index parameter allowing water stress to be a function crop stage
g	= skew coefficient for normal rainfall distribution
h_o	= net solar radiation (ly)

H_i = hydraulic conductivity of soil layer i (mm/h)

HI = harvest index on harvest date

HIAi = harvest index on day i

HIO = harvest index under favorable growing conditions (input)

HU = daily heat units above the crop's base temperature (°C)

HUFH = heat unit index that affects harvest index

HUI = heat unit index

IDA = day of the year

k = decay constant in exponential rainfall distribution (h)

k_c = effective hydraulic conductivity of channel alluvium (mm/h)

k_s = decay constant in sediment settling equation

k_r = decay factor for transmission losses (m/km)

K = MUSLE soil erodibility factor

K_{Ei} = ratio of hydraulic conductivity and vertical equivalent hydraulic conductivity for soil layer i

K_{vi} = vertical equivalent hydraulic conductivity for soil layer i

L = channel length from the most distant point to the subbasin outlet (km)

L_c = average channel flow length (km)

L_{ca} = distance along the channel to the subbasin centroid (km)

LAG = fraction of large aggregates in the detached sediment before deposition (0-1)

LAI = leaf area index--area of plant leaves relative to the soil surface

LAI_{mx} = maximum LAI potential for a crop

LAT = latitude of watershed (degrees)

LC = length of all contours in a subbasin (m)

LG_d = length of grid lines within subbasin in direction d (m)

LS = MUSLE slope length and steepness factor

m = mean daily precipitation (mm)

n = Manning's roughness coefficient

N = number of subbasins

N_d = number of contour crossings for grid lines in direction d

N_g = number of days in growing season

N_l = number of soil layers

ND = number of days in a month

NDD = number of dry days in a month

NRD = number of rainy days in a month

O = percolation rate (mm/d)

P = percolation from soil profile (mm)

P_{in} = inflow volume (m^3)

P_o = threshold volume for a unit channel (m^3)

P(W|D) = Probability of rainfall after a dry day (0-1)

P(W|W) = probability of rainfall after a wet day (0-1)

PAR = photosynthetic active radiation (MJ/m^2)

PCL = fraction of clay in the detached sediment before deposition (0-1)

PCT_j = fraction of particle size j in the detached sediment within the subbasin (0-1)

PE = MUSLE erosion control practice factor

PF = probability factor--assures proper standard deviations of generated weather data

PHU = potential HU for a crop (°C)

POR = total porosity

PR_f = peak rate adjustment factor

PSA = fraction of sand in the detached sediment before deposition (0-1)

PSI = fraction of silt in the detached sediment before deposition (0-1)

PSZ_j = fraction of particle size j in the sediment reaching the subbasin outlet (0-1)

q = flow rate (m^3/s)

q_c = average channel flow rate (m^3/s)

q_p = peak flow rate (m^3/s)

q_{Pl} = peak runoff rate adjusted for transmission losses (m^3/s)

q_s = average surface flow rate (mm/h)

Q = surface runoff volume (mm)

Q_l = surface runoff volume adjusted for transmission losses (mm)

QI = inflow to farm ponds (m^3)

QO = outflow from farm ponds (m^3)

QOR = outflow from reservoirs (m^3)

QR = lateral subsurface flow rate from entire soil profile (mm)

QR_i = lateral subsurface flow rate from soil layer i (mm/d)

r = rainfall intensity (mm/h)

r_{ep} = peak rainfall excess rate (mm/h)

r_p = peak rate of rainfall (mm/h)

R = daily rainfall amount (mm)

$R_{.5}$ = mean maximum 0.5 h rainfall amount per event (mm)

$R_{.5F}$ = maximum 0.5 h rainfall amount for frequency F (mm)

R_t = rainfall amount at time t (mm)

RA = daily solar radiation (ly)

RD = root depth (mm)

REG = minimum crop stress factor

RF = precipitation reduction ratio for subbasins

RR = reservoir principal spillway release rate (m^3/s)

RSD = residue on soil surface (kg/ha)

RWT = fraction of total biomass partitioned to the root system

RZ = maximum root zone depth (mm)

s = SCS curve number retention parameter (mm)

S = land surface slope (m/m)

S_c = channel slope (m/m)

S_d = land surface slope in one grid direction (m/m)

S_L = land surface slope in the length direction (m/m)

S_P = soil temperature damping depth scaling parameter

S_T = land surface slope in the width direction (m/m)

S_w = water surface slope (m/m)

SA = farm pond surface area (ha)

SA_{mx} = maximum farm pond surface area--spillway elevation (ha)

SAG = fraction of small aggregates in the detached sediment before deposition (0-1)

SAN = fraction of primary sand in the original soil mass (0-1)

SC = saturated conductivity (mm/h)

SC_i = saturated conductivity of layer i (mm/h)

SED_T = total sediment reaching the basin outlet (t)

SEP = seepage from ponds and reservoirs (m^3)

SIL = fraction of primary silt in the original soil mass (0-1)

SML = snow melt rate (mm/d)

SNO = water content of snow (mm)

SP = stream power (t/s)

SUT = residue decay factor for soil temperature

SW = soil water content minus 15-bar water content (mm)

SWC = subsurface flow factor for soil water content

t = time (h)

t_c = watershed time of concentration (h)

t_{cc} = time of concentration for channel flow (h)

t_{cs} = time of concentration for surface flow (h)

t_p = hydrograph time to peak (s)

T = daily average air temperature (°C)

T_b = base temperature for a crop (plants start growing) (°C)

T_K = daily average temperature (°K)

T_{mn} = minimum daily air temperature (°C)

T_{mx} = maximum daily air temperature (°C)

T_o = optimal temperature for a crop (°C)

TG = soil surface temperature (°C)

TGB = bare soil surface temperature (°C)

TL = transmission losses (mm)

TS = temperature stress factor for crop growth (0-1)

TT = travel time from subbasin outlet to basin outlet (h)

TT_{ci} = crack flow travel time for layer i (h)

TTi = travel time of flow through soil layer i (h)

TT_{Ri} = subsurface flow time to travel a distance equal the land surface slope length λ in layer i (d)

u_i = plant water use in soil layer i (mm)

u_{pi} = potential plant water use in soil layer i (mm)

U_{RA} = standard deviation of daily solar radiation (ly)

UL = upper limit of soil water storage (porosity minus 15-bar water content (mm)

UT_{mn} = standard deviation of daily minimum air temperature (°C)

UT_{mx} = standard deviation of daily maximum air temperature (°C)

v_c = average channel velocity (m/s)

v_j = fall velocity of sediment particle size j (m/h)

v_s = surface flow velocity (m/s)

V = surface runoff volume (m^3)

VM = volume of water stored in farm ponds (m^3)

VM_{mx} = maximum farm pond volume--spillway elevation (m^3)

VR	= volume of water stored in reservoirs (m^3)
VR_F	= reservoir storage capacity at the emergency spillway crest (m^3)
VR_S	= reservoir storage capacity at the principal spillway crest (m^3)
w	= average channel width of flow (m)
w_1, w_2	= shape parameters for soil water - CN retention parameter relationship
WLV	= weight of crop stover (kg/ha)
WP	= wilting point (15 bar) water content
WS	= water stress factor for crop growth (0-1)
WSYF	= crop parameter expressing drought sensitivity
x, y	= coordinates of rain gages (km)
X	= generated normal deviate
y_c	= channel flow depth (m)
y_j	= distance sediment of particle size d_j will fall during TT (m)
Y	= sediment yield (t)
YD_j	= sediment of particle size j entering the routing reach (t)
YLD	= crop yield (kg/ha)
z	= weighting factor for converting air temperature to soil surface temperature
Z	= soil depth from the surface (mm)
ZM	= distance from bottom of lowest soil layer to the surface (mm)

Appendix II

Weather

Table II.1. Locations where weather generator parameters are available.

Station No .	State	Town	Latitude	Longitude
0001	AL	Birmingham	33.34	86.45
0002	AL	Mobile	30.41	80.15
0003	AL	Montgomery	32.18	86.24
0004	AR	Fort Smith	35.20	94.22
0005	AR	Little Rock	34.73	92.23
0006	AZ	Flagstaff	35.08	111.40
0007	AZ	Phoenix	33.26	112.01
0008	AZ	Yuma	32.67	114.60
0009	CA	Bakersfield	35.42	119.05
0010	CA	Blue Canyon	39.28	120.70
0011	CA	Eureka	40.48	124.10
0012	CA	Fresno	36.46	119.43
0013	CA	Mt. Shasta	41.32	122.32
0014	CA	San Diego	32.44	117.10
0015	CA	San Francisco	37.78	122.42
0016	CO	Colorado Springs	38.49	104.43
0017	CO	Denver	39.77	104.88
0018	CO	Grand Junction	39.07	108.32
0019	CO	Pueblo	38.17	104.31
0020	CT	Hartford (Windsor Locks)	41.93	72.68
0021	DE	Wilmington	39.67	75.60
0022	DC	Washington	38.90	77.05
0023	FL	Jacksonville	30.50	81.70
0024	FL	Miami	25.80	80.27
0025	FL	Tallahassee	30.23	84.22
0026	FL	Tampa	27.58	82.32
0027	GA	Atlanta	33.65	84.43
0028	GA	Agusta	33.22	81.58
0029	GA	Macon	32.42	83.39
0030	GA	Savanah	32.13	81.20
0031	ID	Boise	43.57	116.22
0032	ID	Pocatello	42.55	112.36
0033	IL	Chicago	41.78	87.75
0034	IN	Evansville	38.03	87.32
0035	IN	Fort Wayne	41.00	85.12
0036	IN	Indianapolis	39.73	86.27
0037	IA	Des Moines	41.53	93.65
0038	IA	Dubuque	42.24	90.42
0039	KS	Dodge City	37.46	97.97
0040	KS	Topeka	39.04	95.38
0041	KS	Wichita	37.38	97.25
0042	KY	Lexington	38.02	84.36
0043	KY	Louisville	38.18	85.73
0044	LA	Baton Rouge	30.32	91.09
0045	LA	New Orleans	29.95	90.06

Station No.	State	Town	Latitude	Longitude
0046	LA	Shreveport	32.28	93.49
0047	ME	Caribou	46.52	68.01
0048	ME	Portland	43.65	70.32
0049	MD	Baltimore	39.18	76.67
0050	MA	Boston	42.37	71.03
0051	MA	Nantucket	41.15	70.04
0052	MI	Detroit	42.40	83.00
0053	MI	Grand Rapids	42.53	85.31
0054	MN	Deluth	46.50	92.11
0055	MN	Minneapolis	44.88	93.22
0056	MS	Jackson	32.33	90.22
0057	MO	Columbia	38.58	92.22
0058	MO	Kansas City	39.12	94.60
0059	MO	St. Louis	38.45	90.23
0060	MT	Billings	45.80	108.53
0061	MT	Great Falls	47.29	111.22
0062	MT	Havre	48.33	109.46
0063	MT	Helena	46.36	112.00
0064	MT	Kalispell	48.18	114.16
0065	MT	Miles City	46.26	105.52
0066	NE	Grand Island	40.97	98.32
0067	NE	North Platte	41.08	100.41
0068	NE	Scottsbluff	41.52	103.36
0069	NV	Elko	40.50	115.47
0070	NV	Las Vegas	36.08	115.17
0071	NV	Reno	39.30	119.47
0072	NV	Winnemucca	40.54	117.48
0073	NH	Concord	43.20	71.50
0074	NH	Mt. Washington	44.16	71.18
0075	NJ	Newark	40.70	74.17
0076	NM	Albuqerque	35.05	106.62
0077	NM	Roswell	33.24	104.32
0078	NY	Albany	42.45	73.48
0079	NY	Buffalo	42.93	78.73
0080	NY	New York	40.47	73.58
0081	NY	Syracruse	43.07	76.07
0082	NC	Asheville	35.26	82.32
0083	NC	Greensboro	35.13	80.56
0084	NC	Raleigh	35.87	78.78
0085	ND	Bismarck	46.46	100.45
0086	ND	Williston	48.15	103.62
0087	OH	Cleveland	41.24	81.51
0088	OH	Columbus	40.00	82.88
0089	OH	Toledo	41.36	83.48
0090	OK	Oklahoma City	35.40	97.60

Station No.	State	Town	Latitude	Longitude
0091	OK	Tulsa	36.12	95.54
0092	OR	Burns	42.35	119.03
0093	OR	Meachum	45.30	118.24
0094	OR	Medford	42.22	122.52
0095	OR	Pendleton	45.41	118.51
0096	OR	Portland	45.53	122.67
0097	OR	Salem	44.55	123.01
0098	OR	Sexton Summit	42.37	123.22
0099	PA	Philadelphia	39.53	75.15
0100	PA	Pittsburgh	40.50	80.22
0101	RI	Providence	41.73	71.43
0102	SC	Charleston	32.90	80.03
0103	SC	Columbia	33.57	81.07
0104	SD	Huron	44.23	98.13
0105	SD	Rapid City	44.03	103.05
0106	TN	Chattonooga	35.02	85.12
0107	TN	Knoxville	35.49	83.59
0108	TN	Memphis	35.03	89.59
0109	TN	Nashville	36.12	86.68
0110	TX	Amarillo	35.23	101.70
0111	TX	Austin	30.18	97.42
0112	TX	Brownsville	25.90	97.43
0113	TX	Corpus Christi	27.77	97.50
0114	TX	Dallas	32.85	96.85
0115	TX	El Paso	31.80	106.40
0116	TX	Galveston	29.18	94.48
0117	TX	Houston	29.39	95.17
0118	TX	San Antonio	29.32	98.28
0119	TX	Waco	31.37	97.13
0120	UT	Milford	38.26	113.01
0121	UT	Salt Lake City	40.76	111.97
0122	VA	Norfolk	36.54	76.12
0123	VA	Richmond	37.50	77.33
0124	WA	Olympia	46.58	122.54
0125	WA	Spokane	47.62	117.52
0126	WA	Stampede Pass	47.17	121.20
0127	WA	Walla Walla	46.02	118.20
0128	WA	Yakima	46.34	120.32
0129	WV	Charleston	38.37	81.60
0130	WI	Green Bay	44.29	88.08
0131	WI	La Crosse	43.52	91.15
0132	WI	Madison	43.13	89.33
0133	WI	Milwaukee	42.57	87.54
0134	WY	Cheyenne	41.15	104.82

Table II.2. Weather generator parameters (Station 001 Birmingham, AL).

Avg. monthly values	Jan.	Feb.	Mar.	Apr.	May	June	July	Aug.	Sept.	Oct.	Nov.	Dec.
Maximum temperature (C)	13.1	14.6	18.5	23.3	27.5	31.5	32.3	32.2	03.2	24.9	17.8	13.4
Minimum temperature (C)	1.6	2.8	5.7	9.9	14.2	18.9	20.6	20.1	17.3	10.3	4.4	1.7
Temp. coef. of var.	0.20	0.23	0.16	0.10	0.09	0.07	0.06	0.09	0.08	0.10	0.16	0.18
Solar radiation (Ly)	235	294	349	476	549	548	556	520	442	370	278	204
Max. 5 h rainfall (mm)	21.16	17.78	35.31	21.08	35.56	18.54	27.94	32.77	36.07	19.56	18.29	20.32
Prob. wet after dry	0.264	0.299	0.285	0.245	0.183	0.220	0.307	0.265	0.175	0.144	0.213	0.267
Prob. wet after wet	0.491	0.505	0.475	0.444	0.530	0.581	0.548	0.426	0.480	0.395	0.497	0.495
Mean event rainfall (mm)	11.58	12.45	13.97	13.21	11.43	11.18	10.16	10.67	12.70	11.43	10.67	12.70
Std. dev. event (mm)	15.49	17.02	20.07	16.76	15.24	16.26	12.45	14.99	17.27	15.24	12.70	18.03
Skew coef. event	2.74	3.37	4.28	2.53	2.68	2.67	3.00	2.62	2.20	2.12	1.68	3.07
Latitude (deg)	33.34											
Longitude (deg)	86.45											
Years of record	62											
Elevation (m)	186.0											
10 yr 5 h rainfall (mm)	71.12											
10 yr 6 h rainfall (mm)	152.40											

Appendix III

Hydrology

Table III.1. Runoff curve numbers for hydrologic soil-cover complexes.
(Antecedent moisture condition II, and $I_a = 0.2\ S$)

Land use	Cover: Treatment or practice	Hydrologic condition	Hydrologic soil group A	B	C	D
Fallow	Straight row	----	77	86	91	94
Row crops	Straight row	Poor	72	81	88	91
	"	Good	67	78	85	89
	Contoured	Poor	70	79	84	88
	"	Good	65	65	82	86
	Contoured and terraced	Poor	66	74	80	82
	"	Good	62	71	78	81
Small grain	Straight row	Poor	65	76	84	88
	"	Good	63	75	83	87
	Contoured	Poor	63	74	82	85
	"	Good	61	73	81	84
	Contoured and terraced	Poor	61	72	79	82
	"	Good	59	70	78	81
Close-seeded legumes[1/] or rotation meadow	Straight row	Poor	66	77	85	89
	"	Good	58	72	81	85
	Contoured	Poor	64	75	83	85
	"	Good	55	69	78	83
	Contoured and terraced	Poor	63	73	80	83
	"	Good	51	67	76	80
Pasture or range		Poor	68	79	86	89
		Fair	49	69	79	84
		Good	39	61	74	80
	Contoured	Poor	47	67	81	88
	"	Fair	25	59	75	83
	"	Good	6	35	70	79
Meadow		Good	30	58	71	78
Woods		Poor	45	66	77	83
		Fair	36	60	73	79
		Good	25	55	70	77
Farmsteads		----	59	74	82	86
Roads (dirt)[2/]		----	72	82	87	89
(hard surface)[2/]		----	74	84	90	92

1/ Close-drilled or broadcast.
2/ Including right-of-way.

Taken from USDA (1972).

Table III.1. Continued.

Runoff curve numbers for hydrologic soil-cover complexes in Puerto Rico. (Antecedent moisture condition II, and $I_a = 0.2\,S$).

Cover and condition	Hydrologic soil group A	B	C	D
Fallow	77	86	91	93
Grass (bunch grass or poor stand of sod)	51	70	80	84
Coffee (no ground cover, no terraces)	48	68	79	83
Coffee (with ground cover and terraces)	22	52	68	75
Minor crops (garden or truck crops)	45	66	77	83
Tropical kudzu	19	50	67	74
Sugarcane (trash burned; straight-row)	43	65	77	82
Sugarcane (trash mulch; straight-row)	45	66	77	83
Sugarcane (in holes; on contour)	24	53	69	76
Sugarcane (in furrows; on contour)	32	58	72	79

Runoff curve numbers for hydrologic soil-cover complexes of a typical watershed in Contra Costa County, California. (Antecedent moisture condition II, and $I_a = 0.2\,S$).

Cover	Condition	Hydrologic soil group A	B	C	D
Scrub (native brush)	----	25-30	41-46	57-63	66
Grass-oak (native oaks with understory of forbs and annual grasses)	Good	29-33	43-48	59-65	67
Irrigated pasture	Good	32-37	46-51	62-68	70
Orchard (winter period with understory of cover crop)	Good	37-41	50-55	64-69	71
Range (annual grass)	Fair	46-49	57-60	68-72	74
Small grain (contoured)	Good	61-64	69-71	76-80	81
Truck crops (straight-row)	Good	67-69	74-76	80-83	84
Urban areas:					
Low density (15 to 18 percent impervious surfaces)		69-71	75-78	82-84	86
Medium density (21 to 27 percent impervious surfaces)		71-73	77-80	84-86	88
High density (50 to 75 percent impervious surfaces)		73-75	79-82	86-88	90

Table III.1. Continued.

Runoff curve numbers; tentative estimates for sugarcane hydrology soil-cover complexes in Hawaii. (Antecedent moisture condition II, and I_a = 0.2 S).

Cover and treatment	Hydrologic soil group A	B	C	D
Sugarcane: Limited cover, straight row	67	78	85	89
Partial cover, straight row	49	69	79	84
Complete cover, straight row	39	61	74	80
Limited cover, contoured	65	75	82	86
Partial cover, contoured	25	59	75	83
Complete cover, contoured	6	35	70	79

Limited cover.--Cane newly planted, or ratooned cane with a limited root system; canopy over less than 1/2 the field area.

Partial cover.--Cane in the transition period between limited and complete cover; canopy over 1/2 to nearly the entire field area.

Complete cover.--Cane from the stage of growth when full canopy is provided to the stage at harvest.

Straight row planting is up and down hill or cross-slope on slopes greater than 2 percent. Contoured planting is the usual contouring or cross-slope planting on slopes less than 2 percent.

Table III.1. Continued.

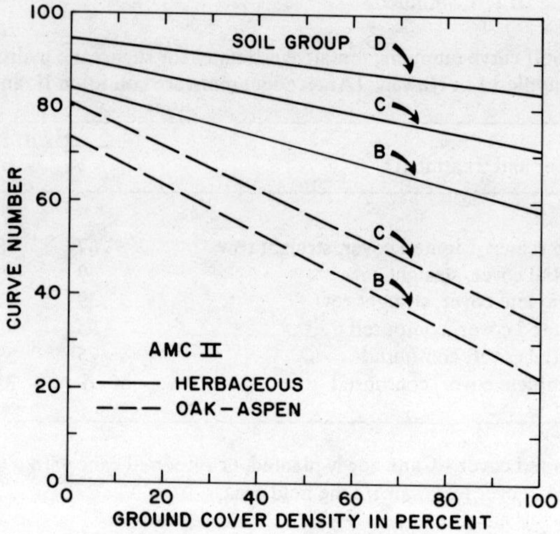

Graph for estimating runoff curve numbers of forest-range complexes in western United States: herbaceous and oak-aspen complexes.

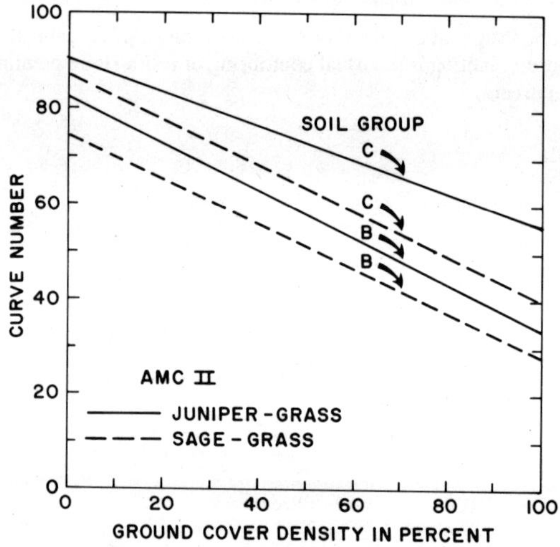

Graph for estimating runoff curve numbers of forest-range complexes in western United States: juniper-grass and sage-grass complexes.

Table III.2. Values of Manning's roughness factor "n".

	Value chosen	Range
I. Channel Flow[1/]		
A. Excavated or dredged		
1. Earth, straight and uniform	0.025	0.016-0.033
2. Earth, winding and sluggish	0.035	0.023-0.05
3. Not maintained, weeds and brush	0.075	0.04-0.14
B. Natural streams		
1. Few trees, stones, or brush	0.05	0.025-0.065
2. Heavy timber and brush	0.10	0.05-0.15
II. Overland Flow[2/]		
Fallow, no residue	0.01	0.008-0.012
Conventional tillage, no residue	0.09	0.06-0.12
Conventional tillage, residue	0.19	0.16-0.22
Chisel plow, no residue	0.09	0.06-0.12
Chisel plow, residue	0.13	0.10-0.16
Fall disking, residue	0.40	0.30-0.50
No till, no residue	0.07	0.04-0.10
No till (0.5 - 1 t/ha^{1})	0.12	0.07-0.17
No till (2 - 9 t/ha^{1})	0.30	0.17-0.47
Rangeland (20% cover)	0.60	
Short grass prairie	0.15	0.10-0.20
Dense grass	0.24	0.17-0.30
Bermudagrass	0.41	0.30-0.48

[1/] Taken from Chow (1959)
[2/] Taken from Engman (1983)

Table III.3. Effective hydraulic conductivity for transmission losses in channel alluvium.

Bed material group	Bed material characteristics	Effective hydraulic conductivity (mm/hr)
1 Very high loss rate	Very clean gravel and large sand. $d_{50} > 2$ mm	>127
2 High loss rate	Clean sand and gravel under field conditions. $d_{50} > 2$ mm	51-127
3 Moderately high loss rate	Sand and gravel mixture with less than a few percent silt-clay.	25-76
4 Moderate loss rate	Mixture of sand and gravel with significant amounts of silt-clay.	6.4-25
5 Very low loss rate	Consolidated bed material with high silt-clay content.	0.025-2.5

Taken from Lane (1982)

Table III.4. Baseflow factor.

Flow characteristics	BFF
1. Perennial streams, flow >75% time	1.0
2. Flow 55 - 75% time	0.75
3. Flow 40 - 55% time	0.50
4. Flow 20 - 40% time	0.25
5. Ephemeral streams	0.00

Taken from Boughton (1984)

Appendix IV

Erosion

Table IV.1. Water erosion control practice factors.

Support Practice Factor (P)

In general, whenever sloping soil is to be cultivated and exposed to erosive rains, the protection offered by sod of close-growing crops in the system needs to be supported by practices that will slow the runoff water and thus reduce the amount of soil it can carry. The most important of these supporting cropland practices are contour tillage, stripcropping on the contour, and terrace systems. Stabilized waterways for the disposal of excess rainfall are a necessary part of each of these practices. By definition, factor P in the USLE is the ratio of soil loss with a specific support practice to the corresponding loss with up-and-down-slope culture. Improved tillage practices, sod-based rotations, fertility treatments, and greater quantities of crop residues left on the field contribute materially to erosion control and frequently provide the major control in a farmer's field. However, these are considered conservation cropping and management practices, and the benefits derived from them are included in C.

P values and slope-length limits for contouring

Land slope (percent)	P value	Maximum length[1] (feet)
1 to 2	0.60	400
3 to 5	0.50	300
6 to 8	0.50	200
9 to 12	0.60	120
13 to 16	0.70	80
17 to 20	0.80	60
21 to 25	0.90	50

[1] Limit may be increased by 25% if residue cover after crop seedlings will regularly exceed 50%.

P values for contour-farmed terraced fields[1]

Land slope (percent)	Farm planning: Contour factor[2]	Farm planning: Stripcrop factor	Computing sediment yield[3]: Graded channels and outlets	Computing sediment yield[3]: Steep backslope underground outlets
1 to 2	0.60	0.30	0.12	0.05
3 to 8	.50	.25	.10	.05
9 to 12	.60	.30	.12	.05
13 to 16	.70	.35	.14	.05
17 to 20	.80	.40	.16	.06
21 to 25	.90	.45	.18	.06

1 Slope length is the horizontal terrace interval. The listed values are for contour farming. No additional contouring factor is used in the computation.

2 Use these values for control if interterrace erosion within specified soil loss tolerances.

3 These values include entrapment efficiency and are used for control of offsite sediment within limits and for estimating the field's contribution to watershed sediment yield.

P values, maximum strip widths, and slope-length limits for contour stripcropping

Land slope (percent)	P values[1] A	P values[1] B	P values[1] C	Strip width[2] (feet)	Maximum length (feet)
1 to 2	0.30	0.45	0.60	130	800
3 to 5	.25	.38	.50	100	600
6 to 8	.25	.38	.50	100	400
9 to 12	.30	.45	.60	80	240
13 to 16	.35	.52	.70	80	160
17 to 20	.40	.60	.80	60	120
21 to 25	.45	.68	.90	50	100

1 P values:

A For 4-year rotation of row crop, small grain with meadow seeding, and 2 years of meadow. A second row crop can replace the small grain if meadow is established in it.

B For 4-year rotation of 2 years row crop, winter grain with meadow seeding, and 1-year meadow.

C For alternate strips of row crop and small grain.

2 Adjust strip-width limit, generally downward, to accommodate widths of farm equipment.

Taken from Wischmeier and Smith (1978)

Appendix V

Soils

Table V.1. Soil-5 parameters [Ripple soil series (USLE K factor .37)].

	Layer					
	1	2	3	4	5	6
Depth (mm)	102	330	508	660	940	1524
Bulk density (t/m^3)	1.30-1.35	1.40-1.45	1.35-1.40	1.40-1.45	1.40-1.45	1.40-1.45
Avail. water cap. (mm/mm)	0.14-0.18	0.14-0.18	0.14-0.18	0.14-0.18	0.17-0.21	0.14-0.18
Sat. conductivity (mm/h)	15.2-50.8	15.2-50.8	15.2-50.8	15.2-50.8	15.2-50.8	15.2-50.8
Clay content (%)	15-20	20-25	25-30	12-17	27-32	30-38
Silt content (%)	25-60	20-55	15-50	18-48	0-28	0-25

Table V.2. Sources of Soils-5 Data.

1. Soils-5 data is included in the computer diskettes. It is accessible from the interactive data entry and automatically entered into a SWRRB data set. Only the data required for SWRRB is given for all 13,788 soil series.

2. Soils-5 can be accessed over commercial phone lines as part of the Environmental Technical Information System (ETIS), which resides on a University of Illinois Computer-the Digital Equipment Corporation's VAX 11/780. Many types of terminals and modems can be used to access this computer. Questions can be directed to:
   ETIS Support Center
   909 West Nevada Street
   Urbana, IL 61801
   (217) 333-1369

3. Hard copy and microfiche of any or all Soils-5 can be obtained from:
   Touchette Corporation
   5701 Enterprise Parkway
   East Syracuse, NY 13057
   (315) 445-0291

4. All necessary Soils-5 data for SWRRB is contained in the SCS County Soil Surveys. All deposition libraries have access to County Soil Surveys.

Table V.3. Mean physical properties of soils.

Texture	Bulk density gm/cm	Total porosity	Volume (in/in) Field capacity 1/3 bar	Wilting point 15 bar	Available water capacity
Coarse sand	1.6	0.40	0.11	0.03	0.08
Sand	1.6	0.40	0.16	0.03	0.13
Fine sand	1.5	0.43	0.18	0.03	0.15
V. fine sand	1.5	0.43	0.27	0.03	0.25
L. coarse sand	1.6	0.40	0.16	0.05	0.11
Loamy sand	1.6	0.40	0.19	0.05	0.14
Loamy f. sand	1.6	0.40	0.22	0.05	0.18
L.v.f. sand	1.6	0.40	0.37	0.05	0.32
Coarse s. loam	1.6	0.40	0.19	0.08	0.11
Sandy loam	1.6	0.40	0.22	0.08	0.14
F. sandy loam	1.7	0.36	0.27	0.08	0.19
V.f. sandy loam	1.6	0.40	0.37	0.08	0.29
Loam	1.6	0.40	0.26	0.11	0.15
Silt loam	1.5	0.43	0.32	0.12	0.20
Silt	1.4	0.47	0.27	0.03	0.24
Sandy clay loam	1.6	0.40	0.30	0.18	0.12
Clay loam	1.6	0.40	0.35	0.22	0.13
Silty clay loam	1.4	0.47	0.36	0.20	0.16
Sandy clay	1.6	0.40	0.28	0.20	0.13
Silty clay	1.5	0.48	0.40	0.30	0.14
Clay	1.4	0.47	0.39	0.28	0.11

Taken from Svetlosanov and Knisel (1982)

Table V.4. SOIL01.DAT

1 AABAB
2 AABERG
3 AARON
4 AASTAD
5 AAZDAHL
6 ABAC
7 ABAJO
8 ABALOBADIAH
9 ABALONE
10 ABARCA
11 ABBAYE
12 ABBEY
13 ABBOTT
14 ABBOTTSTOWN
15 ABCAL
16 ABEGG
17 ABELA
18 ABELL
19 ABERDEEN
20 ABERONE
21 ABERT
22 ABES
23 ABGESE
24 ABILENE
25 ABIQUA
26 ABO
27 ABOR
28 ABRA
29 ABRAHAM
30 ABRAZO
31 ABREU
32 ABSAROKEE
33 ABSCOTA
34 ABSHER
35 ABSTED
36 ABSTED, LOW RAIN
37 ABSTON
38 ACACIO
39 ACADEMY
40 ACADIA
41 ACANA
42 ACANOD
43 ACASCO
44 ACEITUNAS
45 ACEL
46 ACKER
47 ACKERMAN
48 ACKERVILLE
49 ACKETT
50 ACKLEY
51 ACKMEN
52 ACKMORE
53 ACKWATER
54 ACME
55 ACO
56 ACOMA
57 ACORD
58 ACOVE
59 ACREDALE
60 ACREE
61 ACRELANE
62 ACTON
63 ACUFF
64 ACUNA
65 ACY
66 ADA
67 ADAIR
68 ADAMS
69 ADAMSON
70 ADAMSVILLE
71 ADATON
72 ADAVEN
73 ADDICKS
74 ADDIELOU
75 ADE
76 ADEL
77 ADELAIDE
78 ADELANTO
79 ADELINO
80 ADELPHIA
81 ADENA
82 ADGER
83 ADILIS
84 ADJUNTAS
85 ADKINS
86 ADLER
87 ADMAN
88 ADOBE
89 ADOLPH
90 ADOS
91 ADRIAN
92 ADVOKAY
93 AECET
94 AENEAS
95 AERIC CALCIAQUOL
96 AERIC FLUVAQUENT
97 AERIC HALAQUEPTS
98 AERIC HAPLAQUEPT
99 AFTADEN
100 AFTON
101 AGA
102 AGAIPAH
103 AGAN
104 AGAR
105 AGASSIZ
106 AGATE
107 AGATHA
108 AGAWAM
109 AGENCY
110 AGER
111 AGNAL
112 AGNESTON
113 AGNEW
114 AGNOS
115 AGRA
116 AGUA
117 AGUA DULCE
118 AGUA FRIA
119 AGUADILLA
120 AGUALT
121 AGUEDA
122 AGUILARES
123 AGUILITA
124 AGUIRRE
125 AGUSTIN
126 AHL
127 AHLSTROM
128 AHMEEK
129 AHOLT
130 AHREN
131 AHRNKLIN
132 AHTANUM
133 AHWAHNEE
134 AIBONITO
135 AIDO
136 AIKEN
137 AIKMAN
138 AILEY
139 AIMELIIK
140 AINAKEA
141 AINSLEY
142 AINSWORTH
143 AIRMONT
144 AIRPORT
145 AITS
146 AJO
147 AKAKA
148 AKAN
149 AKASKA
150 AKELA
151 AKERCAN
152 AKERUE
153 AKLER
154 ALADDIN
155 ALADSHI
156 ALAE
157 ALAELOA
158 ALAGA
159 ALAKAI
160 ALAMA
161 ALAMANCE
162 ALAMO
163 ALAMOGORDO
164 ALAMOSA
165 ALANOS
166 ALAPAHA
167 ALAPAI
168 ALAZAN
169 ALBAN
170 ALBANO
171 ALBANY
172 ALBATON
173 ALBEE
174 ALBEMARLE
175 ALBERTVILLE
176 ALBINAS
177 ALBINAS, F. SAND
178 ALBION
179 ALBRIGHTS
180 ALBURZ
181 ALBUS
182 ALCESTER
183 ALCOA
184 ALCONA
185 ALCORN
186 ALCOT
187 ALCOVA
188 ALDA
189 ALDAX
190 ALDEN
191 ALDER
192 ALDERDALE
193 ALDERMAND
194 ALDERWOOD
195 ALDI
196 ALDINE
197 ALDING
198 ALDINO
199 ALEDO
200 ALEGROS
201 ALEKNAGIK
202 ALEMEDA
203 ALEX
204 ALEXANDER
205 ALEXANDRIA
206 ALFIR
207 ALFORD
208 ALGANSEE
209 ALGARROBO
210 ALGERITA
211 ALGIERS
212 ALGOA
213 ALGOMA
214 ALHAMBRA
215 ALHARK
216 ALICE
217 ALICEL
218 ALICIA
219 ALIDA
220 ALIKCHI
221 ALINE
222 ALKIRIDGE
223 ALKO
224 ALLAGASH
225 ALLAMORE
226 ALLANTON
227 ALLARD
228 ALLEGHENY
229 ALLEMANDS
230 ALLEN
231 ALLENDALE
232 ALLENS PARK
233 ALLENTINE
234 ALLENWOOD
235 ALLEY
236 ALLHANDS
237 ALLIANCE
238 ALLIGATOR
239 ALLIS
240 ALLISON
241 ALLKER
242 ALLOR
243 ALLOUEZ
244 ALMAC
245 ALMAVILLE
246 ALMENA
247 ALMIRANTE
248 ALMO
249 ALMONT
250 ALMOTA
251 ALMY
252 ALNITE
253 ALO
254 ALOHA
255 ALOMAX
256 ALONA
257 ALONSO
258 ALOVAR
259 ALPENA
260 ALPHA
261 ALPIN
262 ALPON
263 ALPOWA
264 ALROS
265 ALSCO
266 ALSPAUGH
267 ALSTAD
268 ALSTONY
269 ALSUP
270 ALTAMONT
271 ALTAPEAK
272 ALTAVISTA
273 ALTDORF
274 ALTHOUSE
275 ALTICREST
276 ALTITA
277 ALTMAR
278 ALTO
279 ALTOGA
280 ALTON

Appendix V -- 4

281 ALTOONA
282 ALTUDA SERIES
283 ALTURAS
284 ALTUS
285 ALTVAN
286 ALUF
287 ALUM
288 ALUSA
289 ALVARADO
290 ALVIN
291 ALVIRA
292 ALVISO
293 ALVOR
294 ALWILDA
295 ALYAN
296 ALZADA
297 ALZOLA
298 AMADOR
299 AMAGON
300 AMALIA
301 AMALU
302 AMANA
303 AMANDA
304 AMARGOSA
305 AMARILLO
306 AMASA
307 AMBER
308 AMBIA
309 AMBOAT
310 AMBOY
311 AMBRANT
312 AMBRAW
313 AMELIA
314 AMENIA
315 AMENSON
316 AMERICUS
317 AMERY
318 AMES
319 AMESHA
320 AMESMONT
321 AMHERST
322 AMISTAD
323 AMITY
324 AMMON
325 AMOLE
326 AMOR
327 AMOS
328 AMOSTOWN
329 AMPAD
330 AMPHION
331 AMSDEN
332 AMSTERDAM
333 AMTOFT
334 AMWELL
335 AMY
336 ANACAPA
337 ANACOCO
338 ANACONDA
339 ANAHEIM
340 ANAHUAC
341 ANAMITE
342 ANAPRA
343 ANASAZI
344 ANATONE
345 ANAUD
346 ANAVERDE
347 ANAVO
348 ANAWALT
349 ANCHO
350 ANCHOR POINT
351 ANCHORAGE
352 ANCLOTE
353 ANCO
354 ANDAQUEPTS
355 ANDEPTIC UDORTHE
356 ANDERGEORGE
357 ANDERLY
358 ANDERLY, VERY FI
359 ANDERS
360 ANDERSON
361 ANDIC CRYAQUEPT
362 ANDIC CRYOCHREPT
363 ANDIC CRYUMBREPT
364 ANDIC DYSTRIC EU
365 ANDIC HAPLUMBREP
366 ANDIC XEROCHREPT
367 ANDIC XERUMBREPT
368 ANDOK
369 ANDOVER
370 ANDRADA
371 ANDREESON
372 ANDREGG
373 ANDRES
374 ANDREWS
375 ANDRUSIA
376 ANDRY
377 ANED
378 ANETH
379 ANGELICA
380 ANGELINA
381 ANGELO
382 ANGELUS
383 ANGIE
384 ANGLE
385 ANGLEN
386 ANGOLA
387 ANGORA
388 ANGOSTURA
389 ANHALT
390 ANIAK
391 ANIMAS
392 ANINTO
393 ANITA
394 ANKENY
395 ANKLAM
396 ANKONA
397 ANNABELLA
398 ANNANDALE
399 ANNAW
400 ANNEMAINE
401 ANNIS
402 ANNISQUAM
403 ANNISTON
404 ANNONA
405 ANOCON
406 ANOKA
407 ANONES
408 ANSARI
409 ANSEL
410 ANSELMO
411 ANSGAR
412 ANSPING
413 ANT FLAT
414 ANTEL
415 ANTELOPE SPRINGS
416 ANTERO
417 ANTHO
418 ANTHONY
419 ANTIGO
420 ANTILON
421 ANTIOCH
422 ANTLER
423 ANTOINE
424 ANTONITO
425 ANTOSA
426 ANTROBUS
427 ANTWERP
428 ANTY
429 ANUNDE
430 ANVIK
431 ANWAY
432 AOWA
433 APACHE
434 APAKUIE
435 APALACHEE
436 APALO
437 APELDORN
438 APISHAPA
439 APISON
440 APMAT
441 APMAY
442 APOLLO
443 APOPKA
444 APPANOOSE
445 APPERSON
446 APPIAN
447 APPLEBUSH
448 APPLEDELLIA
449 APPLEGATE
450 APPLETON
451 APPLING
452 APRON
453 APT
454 APTAKISIC
455 APTOS
456 AQUIC ARGIBOROLL
457 AQUIC CRYANDEPTS
458 AQUIC HAPLOBOROL
459 AQUIC USTIFLUVEN
460 AQUIC XEROFLUVEN
461 AQUILLA
462 AQUINAS
463 ARABRAB
464 ARADA
465 ARAGON
466 ARAMBURU
467 ARANSAS
468 ARAPAHOE
469 ARAPIEN
470 ARARAT
471 ARAT
472 ARAVAIPA
473 ARAVE
474 ARAVETON
475 ARBELA
476 ARBIDGE
477 ARBOLES
478 ARBONE
479 ARBOR
480 ARBUCKLE
481 ARBURUA
482 ARBUS
483 ARCETTE
484 ARCH
485 ARCHABAL
486 ARCHBOLD
487 ARCHER
488 ARCHERDALE
489 ARCHES
490 ARCHIN
491 ARCHULETA
492 ARCIA
493 ARCO
494 ARCOLA
495 ARD
496 ARDENMONT
497 ARDENVOIR
498 ARDILLA
499 ARDIVEY
500 ARDNAS
501 ARDTOO
502 ARECIBO
503 AREDALE
504 ARENA
505 ARENALES
506 ARENDTSVILLE
507 ARENOSA
508 ARENTS
509 ARENTS, VERY STE
510 ARENTS,SANITARY
511 ARENZVILLE
512 ARGENT
513 ARGENTA
514 ARGIUSTOLLS
515 ARGIXEROLLS
516 ARGONAUT
517 ARGORA
518 ARGYLE
519 ARIEL
520 ARIKARA
521 ARIMO
522 ARIPEKA
523 ARIS
524 ARISPE
525 ARIZO
526 ARKABUTLA
527 ARKANA
528 ARKAQUA
529 ARKONA
530 ARKPORT
531 ARKSON
532 ARKTON
533 ARLAND
534 ARLE
535 ARLINGTON
536 ARLO
537 ARLOVAL
538 ARMAGH
539 ARMCO
540 ARMELLS
541 ARMENIA
542 ARMESA
543 ARMIESBURG
544 ARMIJO
545 ARMINGTON
546 ARMISTEAD
547 ARMITAGE
548 ARMO
549 ARMOINE
550 ARMONA
551 ARMOUR
552 ARMPUP
553 ARMSTER
554 ARMSTRONG
555 ARMUCHEE
556 ARNEGARD
557 ARNHEIM
558 ARNO
559 ARNOLD
560 ARNOT
561 AROL
562 AROSA
563 ARP
564 ARRADA
565 ARRASTRE
566 ARREDONDO
567 ARRIBA
568 ARRINGTON
569 ARRIOLA
570 ARROLIME
571 ARRON
572 ARROWHEAD
573 ARROYADA
574 ARROYO SECO
575 ARSITE
576 ARTA
577 ARTESIA
578 ARTESIAN
579 ARTNOC
580 ARTOIS

581 ARUJO
582 ARUNDEL
583 ARVADA
584 ARVANA
585 ARVESON
586 ARVILLA
587 ARVIN
588 ARZO
589 ASA
590 ASABEAN
591 ASBILL
592 ASCALON
593 ASCAR
594 ASCHOFF
595 ASH SPRINGS
596 ASHBON
597 ASHCROFT
598 ASHDALE
599 ASHDOWN
600 ASHE
601 ASHER
602 ASHFORD
603 ASHFORK
604 ASHGROVE
605 ASHIPPUN
606 ASHKUM
607 ASHLAR
608 ASHLEY
609 ASHLO
610 ASHMED
611 ASHPORT
612 ASHTON
613 ASHUE
614 ASHUELOT
615 ASHWOOD
616 ASKEW
617 ASOTIN
618 ASPARAS
619 ASPEN
620 ASPERMONT
621 ASPERSON
622 ASSININS
623 ASSINNIBOINE
624 ASSUMPTION
625 ASTA
626 ASTATULA
627 ASTOR
628 ASTORIA
629 ATASCO
630 ATASCOSA
631 ATCHEE
632 ATCO
633 ATENCIO
634 ATEPIC
635 ATHELWOLD
636 ATHENA
637 ATHERTON
638 ATHOL
639 ATKINS
640 ATKINSON
641 ATLAS
642 ATLEE
643 ATLOW
644 ATMORE
645 ATOKA
646 ATOMIC
647 ATRAC
648 ATRING
649 ATRYPA
650 ATSION
651 ATTELLA
652 ATTER
653 ATTERBERRY
654 ATTEWAN
655 ATTICA
656 ATTOYAC
657 ATWATER
658 ATWELL
659 ATWOOD
660 AU GRES
661 AUA
662 AUBARQUE
663 AUBBEENAUBBEE
664 AUBERRY
665 AUBREY
666 AUBURN
667 AUBURNDALE
668 AUFCO
669 AUGGIE
670 AUGSBURG
671 AUGUSTA
672 AUGUSTINE
673 AULD
674 AURA
675 AURELIUS
676 AURORA
677 AUSTIN
678 AUSTWELL
679 AUT
680 AUTOMBA
681 AUTRYVILLE
682 AUXVASSE
683 AUZQUI
684 AVA
685 AVALON
686 AVAR
687 AVAWATZ
688 AVENAL
689 AVILLA
690 AVIS
691 AVOCA
692 AVON
693 AVONBURG
694 AVONDA
695 AVONDALE
696 AVONVILLE
697 AVTABLE
698 AWBRIG
699 AXIS
700 AXTELL
701 AYAR
702 AYCOCK
703 AYDELOTTE
704 AYERSVILLE
705 AYLMER
706 AYNOR
707 AYON
708 AYR
709 AYRES
710 AYRSHIRE
711 AYSEES
712 AZAAR
713 AZELTINE
714 AZTALAN
715 AZTEC
716 AZULE
717 AZWELL
718 BAAHISH
719 BABB
720 BABBINGTON
721 BABELTHUAP
722 BACA
723 BACH
724 BACHO
725 BACHUS
726 BACKBAY
727 BACKBONE
728 BACLIFF
729 BACONA
730 BADAXE
731 BADENA
732 BADENAUGH
733 BADGE
734 BADGERTON
735 BADIN
736 BADITO
737 BADLAND
738 BADO
739 BADUS
740 BAGARD
741 BAGDAD
742 BAGGOTT
743 BAGLEY
744 BAHEM
745 BAHIA
746 BAHL
747 BAILE
748 BAILEGAP
749 BAILING
750 BAINVILLE
751 BAIRD HOLLOW
752 BAJURA
753 BAKEOVEN
754 BAKER
755 BAKERSVILLE
756 BALAAM
757 BALCOM
758 BALD
759 BALDER
760 BALDHILL
761 BALDMOUNTAIN
762 BALDOCK
763 BALDWIN
764 BALDY
765 BALE
766 BALLAHACK
767 BALLARD
768 BALLER
769 BALLER VARIANT
770 BALLINGER
771 BALLTOWN
772 BALLY
773 BALM
774 BALMAN
775 BALMLAKE
776 BALMORHEA
777 BALON
778 BALSORA
779 BALTIC
780 BALTIMORE
781 BAMA
782 BAMBER
783 BAMOS
784 BAMTUSH
785 BANAT
786 BANBURY
787 BANCAS
788 BANCROFT
789 BANCY
790 BANDAG
791 BANDERA
792 BANDID
793 BANDON
794 BANE
795 BANGO
796 BANGOR
797 BANGSTON
798 BANIDA
799 BANKARD
800 BANKHEAD
801 BANKS
802 BANLIC
803 BANNEL
804 BANNER
805 BANNING
806 BANNOCK
807 BAPOS
808 BARABOO
809 BARAGA
810 BARATARI
811 BARBAROSA
812 BARBARY
813 BARBERT
814 BARBOUR
815 BARBOURVILLE
816 BARCAVE
817 BARCLAY
818 BARCO
819 BARCUS
820 BARD
821 BARDEN
822 BARDLEY
823 BARELA
824 BARFIELD
825 BARFUSS
826 BARGE
827 BARIO
828 BARISHMAN
829 BARKCAMP
830 BARKERVILLE
831 BARKLEY
832 BARKOF
833 BARLING
834 BARNABE
835 BARNARD
836 BARNES
837 BARNESTON
838 BARNEY
839 BARNHARDT
840 BARNMOT
841 BARNSDALL
842 BARNUM
843 BARODA
844 BAROID
845 BARRADA
846 BARRE
847 BARRETT
848 BARRIER
849 BARRINGTON
850 BARRON
851 BARRONETT
852 BARRY
853 BARSAC
854 BARSHAAD
855 BART
856 BARTINE
857 BARTLE
858 BARTLEY
859 BARTO
860 BARTOME
861 BARTON
862 BARTONFLAT
863 BARVON
864 BARX
865 BASCAL
866 BASCO
867 BASCOM
868 BASCOVY
869 BASEHOR
870 BASHAW
871 BASHER
872 BASILE
873 BASIN
874 BASINGER
875 BASKET
876 BASSEL
877 BASSETT
878 BASSFIELD
879 BASTIAN
880 BASTON

881 BASTROP
882 BASTSIL
883 BATA
884 BATAN
885 BATAVIA
886 BATEMAN
887 BATES
888 BATESVILLE
889 BATH
890 BATTERSON
891 BATTLE CREEK
892 BATZA
893 BAUDETTE
894 BAUER
895 BAUMGARD
896 BAUSCHER
897 BAXENDALE
898 BAXTER
899 BAXTERVILLE
900 BAYAMON
901 BAYARD
902 BAYBORO
903 BAYERTON
904 BAYFIELD
905 BAYLIS
906 BAYMEADE
907 BAYOU
908 BAYOUDAN
909 BAYSHORE
910 BAYSIDE
911 BAYTOWN
912 BAYUCOS
913 BAYVI
914 BAYVIEW
915 BAYWOOD
916 BAZETTE
917 BAZILE
918 BEACH
919 BEACHES
920 BEAD
921 BEADLE
922 BEALAND
923 BEALES
924 BEAM
925 BEAMTON
926 BEANFLAT
927 BEANLAKE
928 BEANO
929 BEAR BASIN
930 BEAR CREEK
931 BEAR LAKE
932 BEAR PRAIRIE
933 BEARDALL
934 BEARDEN
935 BEARDSLEY
936 BEARDSTOWN
937 BEARMOUTH
938 BEARPAW
939 BEARSKIN
940 BEARTRAP
941 BEARVILLE
942 BEARWALLOW
943 BEASLEY
944 BEASON
945 BEATRICE
946 BEAUCOUP
947 BEAUFORD
948 BEAUGHTON
949 BEAUMONT
950 BEAUREGARD
951 BEAUSITE
952 BEAUVAIS
953 BEAVERCREEK
954 BEAVERDAM
955 BEAVERELL
956 BEAVERTON
957 BECKER
958 BECKET
959 BECKLEY
960 BECKMAN
961 BECKS
962 BECKTON
963 BECKVILLE
964 BECKWITH
965 BECKWOURTH
966 BECREEK
967 BEDELL
968 BEDEN
969 BEDFORD
970 BEDINGTON
971 BEDKE
972 BEDNER
973 BEDSTEAD
974 BEDWYR
975 BEE
976 BEEBE
977 BEECHER
978 BEECHWOOD
979 BEEK
980 BEEKMAN
981 BEELEM
982 BEELINE
983 BEEMONT
984 BEENOM
985 BEETVILLE
986 BEGAY
987 BEHANIN
988 BEHEMOTOSH
989 BEHRING
990 BEIGLE
991 BEISIGL
992 BEJE
993 BEJUCOS
994 BELAIN
995 BELATE
996 BELCHER
997 BELDEN
998 BELDING
999 BELEN
1000 BELFAST
1001 BELFIELD
1002 BELFORE
1003 BELGRADE
1004 BELHAVEN
1005 BELINDA
1006 BELJICA
1007 BELK
1008 BELKNAP
1009 BELLAVISTA
1010 BELLE
1011 BELLECHESTER
1012 BELLEHELEN
1013 BELLENMINE
1014 BELLEVILLE
1015 BELLEVUE
1016 BELLICUM
1017 BELLINGHAM
1018 BELLPASS
1019 BELLPINE
1020 BELMEAR
1021 BELMILL
1022 BELMONT
1023 BELMORE
1024 BELPRE
1025 BELTED
1026 BELTON
1027 BELTRAMI
1028 BELTSVILLE
1029 BELUGA
1030 BELVOIR
1031 BELZAR
1032 BEN LOMOND
1033 BENCLARE
1034 BENCO
1035 BENEVOLA
1036 BENEWAH
1037 BENFIELD
1038 BENGAL
1039 BENGE
1040 BENHAM
1041 BENIN
1042 BENITO
1043 BENJAMIN
1044 BENKLIN
1045 BENMAN
1046 BENNDALE
1047 BENNINGTON
1048 BENRIDGE
1049 BENSLEY
1050 BENSON
1051 BENTEEN
1052 BENZ
1053 BEOR
1054 BEOSKA
1055 BEOTIA
1056 BEOWAWE
1057 BEQUINN
1058 BERCUMB
1059 BERDA
1060 BEREA
1061 BERENICETON
1062 BERENT
1063 BERGLAND
1064 BERGQUIST
1065 BERGSTROM
1066 BERGSVIK
1067 BERINO
1068 BERIT
1069 BERKS
1070 BERKSHIRE
1071 BERLAKE
1072 BERLIN
1073 BERMESA
1074 BERMUDIAN
1075 BERNAL
1076 BERNALDO
1077 BERNARD
1078 BERNARDINO
1079 BERNARDSTON
1080 BERNHILL
1081 BERNICE
1082 BERNING
1083 BERNOW
1084 BERRYLAND
1085 BERRYMAN
1086 BERSON
1087 BERTA
1088 BERTAG
1089 BERTELSON
1090 BERTHOUD
1091 BERTIE
1092 BERTOLOTTI
1093 BERTRAM
1094 BERTRAND
1095 BERVILLE
1096 BERWOLF
1097 BERYL
1098 BERZATIC
1099 BESEMAN
1100 BESNER
1101 BESSEMER
1102 BESSIE
1103 BESTROM
1104 BETHANY
1105 BETHERA
1106 BETHESDA
1107 BETIS
1108 BETTERAVIA
1109 BETTS
1110 BEULAH
1111 BEVENT
1112 BEVERIDGE
1113 BEVERLY
1114 BEW
1115 BEWLEYVILLE
1116 BEXAR
1117 BEZO
1118 BEZZANT
1119 BIBB
1120 BICE
1121 BICKERDYKE
1122 BICKLETON
1123 BICKMORE
1124 BICONDOA
1125 BIDDEFORD
1126 BIDDLEMAN
1127 BIDMAN
1128 BIDWELL
1129 BIEBER
1130 BIEDELL
1131 BIENVILLE
1132 BIG BLUE
1133 BIG HORN
1134 BIG TIMBER
1135 BIGARM
1136 BIGBEE
1137 BIGBEND
1138 BIGBROWN
1139 BIGELOW
1140 BIGETTY
1141 BIGFLAT
1142 BIGFORK
1143 BIGNELL
1144 BIGRIVER
1145 BIGSHEEP
1146 BIGSPRING
1147 BIGWIN
1148 BIJOU
1149 BILBO
1150 BILGER
1151 BILLETT
1152 BILLINGS
1153 BILLYCREEK
1154 BILLYHAW
1155 BILTMORE
1156 BIMMER
1157 BINCO
1158 BINDLE
1159 BINFORD
1160 BINGER
1161 BINGHAM
1162 BINGHAMPTON
1163 BINGHAMVILLE
1164 BINNA
1165 BINNSVILLE
1166 BINS
1167 BINTON
1168 BIOYA
1169 BIPPUS
1170 BIRCHBAY
1171 BIRCHFIELD
1172 BIRCHWOOD
1173 BIRDOW
1174 BIRDS
1175 BIRDSALL
1176 BIRDSBORO
1177 BIRDSLEY
1178 BIRDSVIEW
1179 BIRKBECK
1180 BIRMINGHAM

1181 BIRNEY
1182 BIROME
1183 BISBEE
1184 BISCAY
1185 BISGANI
1186 BISHOP
1187 BISMARCK
1188 BISOODI
1189 BISPING
1190 BISSELL
1191 BISSONNET
1192 BIT
1193 BITTER
1194 BITTER SPRING
1195 BITTERROOT
1196 BITTERWATER
1197 BITTON
1198 BIVANS
1199 BIXBY
1200 BIXLER
1201 BJORK
1202 BLACHLY
1203 BLACK BUTTE
1204 BLACK CANYON
1205 BLACK RIDGE
1206 BLACKA
1207 BLACKBURN
1208 BLACKETT
1209 BLACKFOOT
1210 BLACKHALL
1211 BLACKHAMMER
1212 BLACKHAWK
1213 BLACKHOOF
1214 BLACKHORSE
1215 BLACKLEED
1216 BLACKLEG
1217 BLACKLOCK
1218 BLACKMAN
1219 BLACKMOUNT
1220 BLACKNOLL
1221 BLACKOAR
1222 BLACKPIPE
1223 BLACKROCK
1224 BLACKSAN
1225 BLACKSPAR
1226 BLACKSTON
1227 BLACKTHORN
1228 BLACKTOP
1229 BLACKWATER
1230 BLACKWELL
1231 BLADEN
1232 BLAG
1233 BLAGO
1234 BLAINE
1235 BLAIR
1236 BLAIRTON
1237 BLAKABIN
1238 BLAKE
1239 BLAKELAND
1240 BLAKENEY
1241 BLALOCK
1242 BLAMER
1243 BLANCA
1244 BLANCHARD
1245 BLANCHE
1246 BLANCHESTER
1247 BLANCHO
1248 BLANCOT
1249 BLAND
1250 BLANDING
1251 BLANEY
1252 BLANKET
1253 BLANTON
1254 BLANYON
1255 BLASDELL
1256 BLASE
1257 BLASINGAME
1258 BLAYDEN
1259 BLAZBIRD
1260 BLAZON
1261 BLEAKWOOD
1262 BLEDSOE
1263 BLEIBLERVILLE
1264 BLENCOE
1265 BLEND
1266 BLENDON
1267 BLETHEN
1268 BLEVINS
1269 BLEVINTON
1270 BLEWETT
1271 BLICHTON
1272 BLIMO
1273 BLIMSTER
1274 BLINN
1275 BLISS
1276 BLITZEN
1277 BLOCKHOUSE
1278 BLOMFORD
1279 BLOOM
1280 BLOOMFIELD
1281 BLOOMING
1282 BLOOMSDALE
1283 BLOOR
1284 BLOUNT
1285 BLOWERS
1286 BLOWN OUT LAND
1287 BLUCHER
1288 BLUE EARTH
1289 BLUE LAKE
1290 BLUE STAR
1291 BLUEBELL
1292 BLUECHIEF
1293 BLUECREEK
1294 BLUECREST
1295 BLUEDOME
1296 BLUEFLAT
1297 BLUEGROVE
1298 BLUEHILL
1299 BLUEHON
1300 BLUEJOINT
1301 BLUENOSE
1302 BLUEPOINT
1303 BLUERIM
1304 BLUESLIDE
1305 BLUESPRIN
1306 BLUESTONE
1307 BLUEWING
1308 BLUFF
1309 BLUFFDALE
1310 BLUFFTON
1311 BLUFORD
1312 BLUM
1313 BLY
1314 BLYBURG
1315 BLYTHE
1316 BOARDMAN
1317 BOARDTREE
1318 BOASH
1319 BOAZ
1320 BOBBITT
1321 BOBILLO
1322 BOBS
1323 BOBTAIL
1324 BOBTOWN
1325 BOCA
1326 BOCK
1327 BOCKER
1328 BOCKSTON
1329 BODE
1330 BODELL
1331 BODEN
1332 BODENBURG
1333 BODINE
1334 BODORUMPE
1335 BODOT
1336 BOEL
1337 BOELUS
1338 BOERNE
1339 BOESEL
1340 BOETTCHER
1341 BOGAN
1342 BOGART
1343 BOGGS
1344 BOGGY
1345 BOGRAP
1346 BOGUE
1347 BOGUS
1348 BOHANNON
1349 BOHEMIAN
1350 BOHICKET
1351 BOHNA
1352 BOHNLY
1353 BOHNSACK
1354 BOISTFORT
1355 BOJAC
1356 BOJO
1357 BOLAN
1358 BOLAR
1359 BOLD
1360 BOLES
1361 BOLFAR
1362 BOLICKER
1363 BOLIO
1364 BOLIVAR
1365 BOLLING
1366 BOLSA
1367 BOLTON
1368 BOLTUS
1369 BOMAR
1370 BOMBADIL
1371 BOMBAY
1372 BON
1373 BONAIR
1374 BONANZA
1375 BONAPARTE
1376 BONCLAIR
1377 BOND
1378 BONDFARM
1379 BONDMAN
1380 BONDRANCH
1381 BONDUEL
1382 BONE
1383 BONEEK
1384 BONEYARD
1385 BONFIELD
1386 BONFRI
1387 BONG
1388 BONHAM
1389 BONIFAY
1390 BONILLA
1391 BONITA
1392 BONN
1393 BONNEAU
1394 BONNELL
1395 BONNER
1396 BONNERDALE
1397 BONNET
1398 BONNEVILLE
1399 BONNICK
1400 BONNIE
1401 BONNYDOON
1402 BONO
1403 BONSALL
1404 BONTA
1405 BONTI
1406 BONWIER
1407 BOOFORD
1408 BOOFUSS
1409 BOOKCLIFF
1410 BOOKER
1411 BOOKOUT
1412 BOOKWOOD
1413 BOOMER
1414 BOOMTOWN
1415 BOONE
1416 BOONESBORO
1417 BOONTON
1418 BOOTH
1419 BOOTHBAY
1420 BOOTJACK
1421 BOOTS
1422 BOQUILLAS
1423 BORACHO
1424 BORAH
1425 BORDA
1426 BORDEAUX
1427 BORDEN
1428 BORDER
1429 BOREALIS
1430 BORGES
1431 BORIANA
1432 BORKY
1433 BORNSTEDT
1434 BORO
1435 BOROBEY
1436 BOROHEMESTS
1437 BOROHEMISTS
1438 BOROLLIC HAPLARG
1439 BOROLLS
1440 BORREGO
1441 BORSKI
1442 BORTH
1443 BORUP
1444 BORVANT
1445 BOSANKO
1446 BOSCO
1447 BOSKET
1448 BOSLER
1449 BOSQUE
1450 BOSSBURG
1451 BOSTON
1452 BOSTRUM
1453 BOSTWICK
1454 BOSVILLE
1455 BOSWELL
1456 BOSWORTH
1457 BOTELLA
1458 BOTHWELL
1459 BOTHWI
1460 BOTTINEAU
1461 BOTTLE
1462 BOTTLEROCK
1463 BOUFLAT
1464 BOULDER
1465 BOULDER LAKE
1466 BOULDER POINT
1467 BOULDIN
1468 BOULFLAT
1469 BOUNCER
1470 BOUNDARY
1471 BOURBON
1472 BOURNE
1473 BOUSIC
1474 BOW
1475 BOWBAC
1476 BOWBELLS
1477 BOWDISH
1478 BOWDLE
1479 BOWDOIN
1480 BOWDRE

Appendix V -- 8

1481 BOWEN
1482 BOWERS
1483 BOWES
1484 BOWIE
1485 BOWLAKE
1486 BOWMAN
1487 BOWMANSVILLE
1488 BOWNS
1489 BOWSTRING
1490 BOXELDER
1491 BOXFORD
1492 BOXVILLE
1493 BOXWELL
1494 BOY
1495 BOYCE
1496 BOYD
1497 BOYER
1498 BOYKIN
1499 BOYLE
1500 BOYSAG
1501 BOYSEN
1502 BOZE
1503 BOZEMAN
1504 BRACE
1505 BRACEVILLE
1506 BRACKEN
1507 BRACKETT
1508 BRAD
1509 BRADDOCK
1510 BRADEN
1511 BRADENTON
1512 BRADER
1513 BRADSHAW
1514 BRADSON
1515 BRADWAY
1516 BRADY
1517 BRADYVILLE
1518 BRAGG
1519 BRAHAM
1520 BRAILSFORD
1521 BRAINERD
1522 BRALLIER
1523 BRAM
1524 BRAMARD
1525 BRAMLETT
1526 BRAMWELL
1527 BRANCH
1528 BRAND
1529 BRANDENBURG
1530 BRANDON
1531 BRANDYWINE
1532 BRANFORD
1533 BRANHAM
1534 BRANTFORD
1535 BRANTLEY
1536 BRANYON
1537 BRASHEAR
1538 BRASSFIELD
1539 BRATTON
1540 BRAUN
1541 BRAVANE
1542 BRAXTON
1543 BRAY
1544 BRAYTON
1545 BRAZITO
1546 BRAZON
1547 BRAZORIA
1548 BRECKENRIDGE
1549 BRECKNOCK
1550 BRECKSVILLE
1551 BREECE
1552 BREGAR
1553 BREIEN
1554 BREKO
1555 BREMER
1556 BREMO
1557 BREMS
1558 BRENDA
1559 BRENHAM
1560 BRENNAN
1561 BRENNER
1562 BRENT
1563 BRENTON
1564 BRENTSVILLE
1565 BRENTWOOD
1566 BRESSA
1567 BRESSER
1568 BREVARD
1569 BREVORT
1570 BREW
1571 BREWER
1572 BREWSTER
1573 BREWTON
1574 BRIBUTTE
1575 BRICKEL
1576 BRICKMILL
1577 BRICKTON
1578 BRICO
1579 BRIDGE
1580 BRIDGECREEK
1581 BRIDGEHAMPTON
1582 BRIDGEPORT
1583 BRIDGER
1584 BRIDGESON
1585 BRIDGET
1586 BRIEDWELL
1587 BRIEF
1588 BRIGGS
1589 BRIGGSDALE
1590 BRIGGSVILLE
1591 BRIGHTON
1592 BRIGHTWOOD
1593 BRILEY
1594 BRILL
1595 BRILLIANT
1596 BRIMFIELD
1597 BRIMLEY
1598 BRIMSTONE
1599 BRINEGAR
1600 BRINKERT
1601 BRINKERTON
1602 BRINNUM
1603 BRIONES
1604 BRIOS
1605 BRISBANE
1606 BRISCOT
1607 BRISKY
1608 BRISTOW
1609 BRITTO
1610 BRITWATER
1611 BROAD
1612 BROAD CANYON
1613 BROADALBIN
1614 BROADAX
1615 BROADBROOK
1616 BROADHEAD
1617 BROADHURST
1618 BROADMOOR
1619 BROADUS
1620 BROADWELL
1621 BROCK
1622 BROCKET
1623 BROCKLISS
1624 BROCKMAN
1625 BROCKO
1626 BROCKPORT
1627 BROCKROAD
1628 BROCKSBURG
1629 BROCKTON
1630 BROCKWAY
1631 BROCKWELL
1632 BRODALE
1633 BRODY
1634 BROE
1635 BROGAN
1636 BROGDON
1637 BROKAW
1638 BROLLIAR
1639 BROMER
1640 BROMIDE
1641 BROMO
1642 BRONAUGH
1643 BRONCHO
1644 BRONSON
1645 BRONTE
1646 BROOKE
1647 BROOKFIELD
1648 BROOKINGS
1649 BROOKLYN
1650 BROOKMAN
1651 BROOKSHIRE
1652 BROOKSIDE
1653 BROOKSTON
1654 BROOKSVILLE
1655 BROOME
1656 BROPHY
1657 BROSE
1658 BROSELEY
1659 BROSS
1660 BROUGHTON
1661 BROWARD
1662 BROWER
1663 BROWNDELL
1664 BROWNELL
1665 BROWNFIELD
1666 BROWNLEE
1667 BROWNRIGG
1668 BROWNSCOMBE
1669 BROWNSDALE
1670 BROWNSTO
1671 BROWNSVILLE
1672 BROWNTON
1673 BROXON
1674 BROYLES
1675 BRUBECK
1676 BRUCE
1677 BRUFFY
1678 BRUIN
1679 BRUNCAN
1680 BRUNDAGE
1681 BRUNEEL
1682 BRUNELDA
1683 BRUNO
1684 BRUNSWICK
1685 BRUNZELL
1686 BRUSSELS
1687 BRUSSETT
1688 BRYAN
1689 BRYANT
1690 BRYARLY
1691 BRYCAN
1692 BRYCE
1693 BRYMAN
1694 BRYSTAL
1695 BUB
1696 BUBUS
1697 BUCAN
1698 BUCHANAN
1699 BUCHEL
1700 BUCHENAU
1701 BUCKAROO
1702 BUCKBAY
1703 BUCKCREEK
1704 BUCKEYE
1705 BUCKHALL
1706 BUCKHOUSE
1707 BUCKLAND
1708 BUCKLE
1709 BUCKLEBAR
1710 BUCKLEY
1711 BUCKLICK
1712 BUCKLON
1713 BUCKNELL
1714 BUCKNEY
1715 BUCKPEAK
1716 BUCKS
1717 BUCKSKIN
1718 BUCKTON
1719 BUDE
1720 BUDIHOL
1721 BUDLEWIS
1722 BUELL
1723 BUENA VISTA
1724 BUFFARAN
1725 BUFFINGTON
1726 BUFFMEYER
1727 BUFFORK
1728 BUFTON
1729 BUHRIG
1730 BUICK
1731 BUIST
1732 BUKO
1733 BUKREEK
1734 BULAKE
1735 BULKLEY
1736 BULL RUN
1737 BULL TRAIL
1738 BULLARDS
1739 BULLCREEK
1740 BULLFAT
1741 BULLION
1742 BULLNEL
1743 BULLOCK
1744 BULLREY
1745 BULLUMP
1746 BULLWINKLE
1747 BULLY
1748 BULOW
1749 BULRING
1750 BUNCOMBE
1751 BUNDO
1752 BUNDORF
1753 BUNDY
1754 BUNDYMAN
1755 BUNEJUG
1756 BUNKER
1757 BUNKERHILL
1758 BUNKY
1759 BUNNELL
1760 BUNTINGVILLE
1761 BUNYAN
1762 BURBANK
1763 BURCH
1764 BURCHAM
1765 BURCHARD
1766 BURDETT
1767 BUREN
1768 BURGESS
1769 BURGI
1770 BURKE
1771 BURKETOWN
1772 BURKEVILLE
1773 BURKHARDT
1774 BURLEIGH
1775 BURLESON
1776 BURLEWASH
1777 BURLINGTON
1778 BURMAH
1779 BURNAC
1780 BURNBOROUGH

1781 BURNEL
1782 BURNETTE
1783 BURNHAM
1784 BURNSIDE
1785 BURNSVILLE
1786 BURNT LAKE
1787 BURR
1788 BURRITA
1789 BURROWSVILLE
1790 BURSLEY
1791 BURSON
1792 BURT
1793 BURTON
1794 BURWELL
1795 BUSBY
1796 BUSE
1797 BUSHER
1798 BUSHMAN
1799 BUSHNELL
1800 BUSHVALLEY
1801 BUSKA
1802 BUSSY
1803 BUSTER
1804 BUSTI
1805 BUSYWILD
1806 BUTCHE
1807 BUTLER
1808 BUTLERTOWN
1809 BUTTERFIELD
1810 BUTTERS
1811 BUTTON
1812 BUTTONWILLOW
1813 BUXIN
1814 BUXTON
1815 BUZZN
1816 BYARS
1817 BYBEE
1818 BYLER
1819 BYLO
1820 BYNUM
1821 BYRAM
1822 BYRNIE
1823 CABALLO
1824 CABARTON
1825 CABBA
1826 CABBART
1827 CABEZON
1828 CABIN
1829 CABINET
1830 CABLE
1831 CABO ROJO
1832 CABOOSE
1833 CABOT
1834 CABSTON
1835 CACHE
1836 CACIQUE
1837 CACTUSFLAT
1838 CADDO
1839 CADEVILLE
1840 CADILLAC
1841 CADIZ
1842 CADMUS
1843 CADOMA
1844 CAGEY
1845 CAGLE
1846 CAGUABO
1847 CAGWIN
1848 CAHABA
1849 CAHONA
1850 CAID
1851 CAINHOY
1852 CAIRO
1853 CAJALCO
1854 CAJETE
1855 CAJON
1856 CALABASAS
1857 CALAMINE
1858 CALAPOOYA
1859 CALAVERAS
1860 CALAWAH
1861 CALCIORTHIDS
1862 CALCO
1863 CALCOUSTA
1864 CALCROSS
1865 CALD
1866 CALDER
1867 CALDWELL
1868 CALE
1869 CALEAST
1870 CALEB
1871 CALEDONIA
1872 CALENDAR
1873 CALHI
1874 CALHOUN
1875 CALICO
1876 CALICOTT
1877 CALIFON
1878 CALIMUS
1879 CALITA
1880 CALIZA
1881 CALKINS
1882 CALLABO
1883 CALLAHAN
1884 CALLAN
1885 CALLEGUAS
1886 CALLINGS
1887 CALLISBURG
1888 CALLOWAY
1889 CALMAR
1890 CALODO
1891 CALOOSA
1892 CALOUSE
1893 CALPAC
1894 CALPEAK
1895 CALPINE
1896 CALROY
1897 CALVERTON
1898 CALVIN
1899 CALVISTA
1900 CALWOODS
1901 CAMAGUEY
1902 CAMARGO
1903 CAMARILLO
1904 CAMAS
1905 CAMATTA
1906 CAMBARGE
1907 CAMBERN
1908 CAMBERT
1909 CAMBETH
1910 CAMBORTHIDS
1911 CAMBRIA
1912 CAMBRIDGE
1913 CAMDEN
1914 CAMELBACK
1915 CAMERON
1916 CAMILLUS
1917 CAMINO
1918 CAMPBELL
1919 CAMPBELLTON
1920 CAMPIA
1921 CAMPO
1922 CAMPONE
1923 CAMPSPASS
1924 CAMPUS
1925 CAMRODEN
1926 CANA
1927 CANAAN
1928 CANADIAN
1929 CANADICE
1930 CANALOU
1931 CANANDAIGUA
1932 CANASERAGA
1933 CANAVERAL
1934 CANBURN
1935 CANDELARIA
1936 CANDELERO
1937 CANDERLY
1938 CANDLER
1939 CANDOR
1940 CANE
1941 CANEADEA
1942 CANEEK
1943 CANELO
1944 CANEST
1945 CANEYVILLE
1946 CANEZ
1947 CANFIELD
1948 CANISTEO
1949 CANLON
1950 CANNELL
1951 CANNING
1952 CANNON
1953 CANOE
1954 CANOVA
1955 CANTALA
1956 CANTEY
1957 CANTON
1958 CANTON BEND
1959 CANTRIL
1960 CANTUA
1961 CANTUCHE
1962 CANUTIO
1963 CANYON
1964 CAPAC
1965 CAPAY
1966 CAPE
1967 CAPE FEAR
1968 CAPEHORN
1969 CAPERS
1970 CAPERTON
1971 CAPHOR
1972 CAPILLO
1973 CAPISTRANO
1974 CAPITAN
1975 CAPJAC
1976 CAPLES
1977 CAPONA
1978 CAPPS
1979 CAPSHAW
1980 CAPTINA
1981 CAPTIVA
1982 CAPULIN
1983 CARACOLES
1984 CARADAN
1985 CARALAMPI
1986 CARBENGLE
1987 CARBO
1988 CARBOL
1989 CARBONDALE
1990 CARCITY
1991 CARDIFF
1992 CARDIGAN
1993 CARDINGTON
1994 CARDON
1995 CAREFREE
1996 CAREY
1997 CAREY LAKE
1998 CARGILL
1999 CARIBEL
2000 CARIBOU
2001 CARIOCA
2002 CARJO
2003 CARLIN
2004 CARLINTON
2005 CARLISLE
2006 CARLITO
2007 CARLOS
2008 CARLOW
2009 CARLSBAD
2010 CARLSBORG
2011 CARLSON
2012 CARLSTROM
2013 CARLTON
2014 CARMACK
2015 CARMEL
2016 CARMI
2017 CARMICHAEL
2018 CARMODY
2019 CARNAGE
2020 CARNAHAN
2021 CARNASAW
2022 CARNEGIE
2023 CARNERO
2024 CARNEY
2025 CAROLINE
2026 CAROLLO
2027 CARON
2028 CARPENTER
2029 CARR
2030 CARRACAS
2031 CARRANZA
2032 CARRICK
2033 CARRIZALES
2034 CARRIZO
2035 CARRYBACK
2036 CARSITAS
2037 CARSON
2038 CARSTAIRS
2039 CARSTUMP
2040 CART
2041 CARTAGENA
2042 CARTECAY
2043 CARTER
2044 CARTERET
2045 CARTHAGE
2046 CARUSO
2047 CARUTHERSVILLE
2048 CARVER
2049 CARWILE
2050 CARYTOWN
2051 CARYVILLE
2052 CASA GRANDE
2053 CASABONNE
2054 CASAGA
2055 CASCADE
2056 CASCAJO
2057 CASCILLA
2058 CASCO
2059 CASE
2060 CASEY
2061 CASHEL
2062 CASHION
2063 CASHMERE
2064 CASHMONT
2065 CASITO
2066 CASMOS
2067 CASPAR
2068 CASPIANA
2069 CASS
2070 CASSIA
2071 CASSIRO
2072 CASSOLARY
2073 CASTAIC
2074 CASTALIA
2075 CASTANA
2076 CASTELL
2077 CASTELLEIA
2078 CASTELLEIA, DARK
2079 CASTELLO
2080 CASTEPHEN

2081 CASTILE
2082 CASTINO
2083 CASTLE
2084 CASTLEVALE
2085 CASTNER
2086 CASTO
2087 CASTON
2088 CASTRO
2089 CASTROVILLE
2090 CASUSE
2091 CASWELL
2092 CATALINA
2093 CATALPA
2094 CATAMOUNT
2095 CATANO
2096 CATARACT
2097 CATARINA
2098 CATASKA
2099 CATAULA
2100 CATCHELL
2101 CATELLI
2102 CATERL SERIES
2103 CATH
2104 CATHARPIN
2105 CATHAY
2106 CATHCART
2107 CATHEDRAL
2108 CATHERINE
2109 CATHLAMET
2110 CATHRO
2111 CATILLA
2112 CATLA
2113 CATLETT
2114 CATLIN
2115 CATMAN
2116 CATNIP
2117 CATOCTIN
2118 CATOOSA
2119 CATPOINT
2120 CATTCREEK
2121 CATTO
2122 CAUDLE
2123 CAUSEWA
2124 CAUSEY
2125 CAVAL
2126 CAVANAUGH
2127 CAVE
2128 CAVEHILL
2129 CAVELT
2130 CAVENDISH
2131 CAVO
2132 CAVODE
2133 CAVOUR
2134 CAYA
2135 CAYAGUA
2136 CAYUGA
2137 CAYUSE
2138 CAZADERO
2139 CAZENOVIA
2140 CEBOLIA
2141 CEBONE
2142 CEBOYA
2143 CECIL
2144 CEDA
2145 CEDAR BUTTE
2146 CEDAR MOUNTAIN
2147 CEDARAN
2148 CEDARBLUFF
2149 CEDARGAP
2150 CEDARHILL
2151 CEDARPASS
2152 CEDONIA
2153 CELACY
2154 CELESTE
2155 CELETON
2156 CELINA
2157 CELIO
2158 CELLAR
2159 CELSOSPRINGS
2160 CEMBER
2161 CENCOVE
2162 CENIZA
2163 CENTENARY
2164 CENTER
2165 CENTER CREEK
2166 CENTERBURG
2167 CENTERFIELD
2168 CENTERVILLE
2169 CENTISSIMA
2170 CENTRAL POINT
2171 CENTRALIA
2172 CERBAT
2173 CERESCO
2174 CERLIN
2175 CERRILLOS
2176 CERRO
2177 CESTNIK
2178 CETRACK
2179 CHACON
2180 CHAD
2181 CHAFFEE
2182 CHAGRIN
2183 CHAIRES
2184 CHAIX
2185 CHALCO
2186 CHALFONT
2187 CHALMERS
2188 CHAMA
2189 CHAMATE
2190 CHAMBERINO
2191 CHAMISE
2192 CHAMITA
2193 CHAMOKANE
2194 CHAMPAGNE
2195 CHAMPION
2196 CHANAC
2197 CHANCE
2198 CHANCELLOR
2199 CHANDLER
2200 CHANEY
2201 CHANNAHON
2202 CHANNING
2203 CHANTA
2204 CHANTIER
2205 CHAPANOKE
2206 CHAPERTON
2207 CHAPIN
2208 CHAPIN VARIANT
2209 CHAPMAN
2210 CHAPOT
2211 CHAPPELL
2212 CHARCO
2213 CHARCOL
2214 CHARD
2215 CHARDOTON
2216 CHARETTE
2217 CHARGO
2218 CHARITON
2219 CHARLDS
2220 CHARLEBOIS
2221 CHARLES
2222 CHARLESTON
2223 CHARLEVOIX
2224 CHARLOS
2225 CHARLOTTE
2226 CHARLTON
2227 CHARNOCK
2228 CHASE
2229 CHASEBURG
2230 CHASEVILLE
2231 CHASKA
2232 CHASTAIN
2233 CHATBURN
2234 CHATCOLET
2235 CHATEAU
2236 CHATFIELD
2237 CHATHAM
2238 CHATSWORTH
2239 CHATT
2240 CHATUGE
2241 CHAUMONT
2242 CHAUNCEY
2243 CHAUTAUQUA
2244 CHAVIES
2245 CHAWANAKEE
2246 CHAYSON
2247 CHAZOS
2248 CHEADLE
2249 CHEAHA
2250 CHECHI
2251 CHECKETT
2252 CHEDEHAP
2253 CHEDSEY
2254 CHEEBE
2255 CHEEKTOWAGA
2256 CHEESEMAN
2257 CHEHALEM
2258 CHEHALIS
2259 CHEHULPUM
2260 CHELAN
2261 CHELSEA
2262 CHEMAWA
2263 CHEN
2264 CHENA
2265 CHENANGO
2266 CHENAULT
2267 CHENEGA
2268 CHENEY
2269 CHENNEBY
2270 CHENOWETH
2271 CHEQUEST
2272 CHERIONI
2273 CHEROKEE
2274 CHERRY
2275 CHERRY SPRING
2276 CHERRYHILL
2277 CHERUM
2278 CHESAW
2279 CHESHIRE
2280 CHESHNINA
2281 CHESTATEE
2282 CHESTER
2283 CHESTERTON
2284 CHESTONIA
2285 CHESUNCOOK
2286 CHETCO
2287 CHETEK
2288 CHETWYND
2289 CHEVAL
2290 CHEVELON
2291 CHEVIOT
2292 CHEWACLA
2293 CHEWELAH
2294 CHEYENNE
2295 CHIA
2296 CHIARA
2297 CHICANE
2298 CHICKAHOMINY
2299 CHICKASHA
2300 CHICKREEK
2301 CHIEFLAND
2302 CHIGLEY
2303 CHIKAMIN
2304 CHILAO
2305 CHILCOTT
2306 CHILGREN
2307 CHILHOWIE
2308 CHILI
2309 CHILICOTAL
2310 CHILL
2311 CHILLUM
2312 CHILMARK
2313 CHILOQUIN
2314 CHILPER
2315 CHILSON
2316 CHILTON
2317 CHIMAYO
2318 CHIME
2319 CHIMENEA
2320 CHINAPOINT
2321 CHINCHALLO
2322 CHINIAK
2323 CHINO
2324 CHINOOK
2325 CHIPETA
2326 CHIPLEY
2327 CHIPMAN
2328 CHIPOLA
2329 CHIPPENY
2330 CHIPPEWA
2331 CHIRENO
2332 CHIRICAHUA
2333 CHISCA
2334 CHISMORE
2335 CHISOLM
2336 CHISPA
2337 CHITINA
2338 CHITTUM
2339 CHITWOOD
2340 CHIVATO
2341 CHIWAUKUM
2342 CHIWAWA
2343 CHLORIDE
2344 CHO
2345 CHOATES
2346 CHOBEE
2347 CHOCCOLOCCO
2348 CHOCK
2349 CHOCORUA
2350 CHOICE
2351 CHOOP
2352 CHOPTIE
2353 CHORALMONT
2354 CHOSKA
2355 CHOTEAU
2356 CHOWAN
2357 CHRIS
2358 CHRISMAN
2359 CHRISTIAN
2360 CHRISTIANA
2361 CHRISTIANBURG
2362 CHRISTINE
2363 CHRISTOFF
2364 CHRISTY
2365 CHROME
2366 CHRYSLER
2367 CHUALAR
2368 CHUBBS
2369 CHUCKANUT
2370 CHUCKAWALLA
2371 CHUCKLES
2372 CHUCKRIDGE
2373 CHUGCREEK
2374 CHUGTER
2375 CHULITNA
2376 CHUMALL
2377 CHUMSTICK
2378 CHUPADERA
2379 CHURCH
2380 CHURCHILL

2381 CHURCHVILLE
2382 CHURN
2383 CHUSKA
2384 CHUTE
2385 CIALES
2386 CIBEQUE
2387 CIBO
2388 CIBOLA
2389 CIDRAL
2390 CIENEBA
2391 CIENO
2392 CIFIC
2393 CIMARRON
2394 CINCINNATI
2395 CINCO
2396 CINDER LAND
2397 CINDERHURST
2398 CINEBAR
2399 CINNAMON
2400 CINTRONA
2401 CIPRIANO
2402 CIRCLEBACK
2403 CIRCLEBAR
2404 CIRCLEVILLE
2405 CISCO
2406 CISNE
2407 CISPUS
2408 CITADEL
2409 CITICO
2410 CLACKAMAS
2411 CLAIBORNE
2412 CLAIRE
2413 CLAIREMONT
2414 CLAKELEN
2415 CLALLAM
2416 CLAM GULCH
2417 CLAMO
2418 CLANA
2419 CLANALPINE
2420 CLANTON
2421 CLAPPER
2422 CLAREMORE
2423 CLARENCE
2424 CLARENDON
2425 CLARESON
2426 CLAREVILLE
2427 CLARINDA
2428 CLARION
2429 CLARITA
2430 CLARK
2431 CLARK FORK
2432 CLARKELEN
2433 CLARKRANGE
2434 CLARKSBURG
2435 CLARKSDALE
2436 CLARKSVILLE
2437 CLARNO
2438 CLATO
2439 CLATSOP
2440 CLAVERACK
2441 CLAVICON
2442 CLAWSON
2443 CLAYBURN
2444 CLAYSPRINGS
2445 CLAYTON
2446 CLE ELUM
2447 CLEAR LAKE
2448 CLEARBROOK
2449 CLEARFIELD
2450 CLEARFORK
2451 CLEARWATER
2452 CLEAVAGE
2453 CLEAVER
2454 CLEBIT
2455 CLEGG
2456 CLEGHORN
2457 CLEMAN
2458 CLEMENTINE
2459 CLEMS
2460 CLEMVILLE
2461 CLENDENEN
2462 CLEORA
2463 CLERF
2464 CLERGERN
2465 CLERMONT
2466 CLEVELAND
2467 CLEVERLY
2468 CLICK
2469 CLIFFDELL
2470 CLIFFDOWN
2471 CLIFFHOUSE
2472 CLIFFORD
2473 CLIFFSAND
2474 CLIFTERSON
2475 CLIFTON
2476 CLIFTY
2477 CLIMARA
2478 CLIMAX
2479 CLIME
2480 CLINETOP
2481 CLINT
2482 CLINTON
2483 CLODINE
2484 CLONTARF
2485 CLOQUALLUM
2486 CLOQUATO
2487 CLOQUET
2488 CLOSKEY
2489 CLOTHO
2490 CLOUD PEAK
2491 CLOUD RIM
2492 CLOUDCROFT
2493 CLOUDLAND
2494 CLOUGH
2495 CLOVELLY
2496 CLOVER SPRINGS
2497 CLOVERDALE
2498 CLOVERLAND
2499 CLOVIS
2500 CLOWERS
2501 CLOWFIN
2502 CLUFF
2503 CLUNIE
2504 CLURDE
2505 CLURO
2506 CLYDE
2507 CLYMER

SOIL02.DAT

1 COACHELLA
2 COAHUILA
3 COAL CREEK
4 COALBANK
5 COALDRAW
6 COALMONT
7 COAMO
8 COARSEGOLD
9 COASTAL BEACH
10 COASTAL BEACHES
11 COASTAL DUNES
12 COATSBURG
13 COBAT
14 COBB
15 COBBSFORK
16 COBEN
17 COBERLY
18 COBEY
19 COBOC
20 COBURG
21 COCHETOPA
22 COCHINA
23 COCHITI
24 COCOA
25 COCODRIE
26 COCOLALLA
27 CODLEY
28 CODORUS
29 COE
30 COEROCK
31 COFF
32 COFFEEN
33 COGGON
34 COGNA
35 COGSWELL
36 COHAGEN
37 COHASSET
38 COHOCTAH
39 COHOE
40 COILS
41 COIT
42 COKEDALE
43 COKEL
44 COKER
45 COKESBURY
46 COKEVILLE
47 COLAND
48 COLBAR
49 COLBERT
50 COLBURN
51 COLBY
52 COLDCREEK
53 COLE
54 COLEMAN
55 COLEMANTOWN
56 COLESTINE
57 COLFAX
58 COLHILL
59 COLIBRO
60 COLINAS
61 COLITA
62 COLLAMER
63 COLLARD
64 COLLAYOMI
65 COLLBRAN
66 COLLEGEDALE
67 COLLEGIATE
68 COLLETT
69 COLLIER
70 COLLINGTON
71 COLLINS
72 COLLINSTON
73 COLLINSVILLE
74 COLLINWOOD
75 COLMA
76 COLMOR
77 COLNEVEE
78 COLO
79 COLOCKUM
80 COLOMA
81 COLOMBO
82 COLONA
83 COLONIE
84 COLONVILLE
85 COLORADO
86 COLOROCK
87 COLOROW
88 COLOSO
89 COLOSSE
90 COLOTE
91 COLP
92 COLRAIN
93 COLSAVAGE
94 COLTER
95 COLTHORP
96 COLTON
97 COLTS NECK
98 COLUMBIA
99 COLUMBINE
100 COLUMBUS
101 COLUSA
102 COLVARD
103 COLVILLE
104 COLVIN
105 COLWOOD
106 COLY
107 COLYER
108 COMAD
109 COMAR
110 COMBE
111 COMBS
112 COMER
113 COMETA
114 COMFORT
115 COMFREY
116 COMITAS
117 COMLY
118 COMMERCE
119 COMMSKI
120 COMO
121 COMOBABI
122 COMODORE
123 COMORO
124 COMPASS
125 COMPTCHE
126 COMSTOCK
127 COMUS
128 CONA
129 CONABY
130 CONALB
131 CONANT
132 CONASAUGA
133 CONATA
134 CONBOY
135 CONCEPCION
136 CONCHAS
137 CONCHO
138 CONCONULLY
139 CONCORD
140 CONDA
141 CONDIE
142 CONDIT
143 CONDON
144 CONE
145 CONECUH
146 CONEJO
147 CONESTOGA
148 CONESUS
149 CONETOE
150 CONGAREE
151 CONGER
152 CONGLE
153 CONI
154 CONIC
155 CONKLIN
156 CONLEN
157 CONLEY
158 CONNEAUT
159 CONNEL
160 CONNERTON
161 CONOSTA
162 CONOTTON
163 CONOVER
164 CONOWINGO
165 CONPEAK
166 CONRAD
167 CONROE
168 CONSEJO
169 CONSER
170 CONSTABLE
171 CONSTANCIA
172 CONSUMO
173 CONTEE
174 CONTIDE
175 CONTINE
176 CONTINENTAL
177 CONTO
178 CONTRA COSTA
179 CONTRARY
180 CONVENT
181 COOERS
182 COOK
183 COOKPORT
184 COOLBRITH
185 COOLIDGE
186 COOLVILLE
187 COOMBS
188 COOPER
189 COOSAW
190 COOSBAY
191 COOTER
192 COPAKE
193 COPALIS
194 COPASTON
195 COPELAND
196 COPEMAN
197 COPENHAGEN
198 COPITA
199 COPPER RIVER
200 COPPEREID
201 COPPERTON
202 COPPOCK
203 COPSEY
204 COQUAT
205 COQUILLE
206 CORA
207 CORAL
208 CORALLAKE
209 CORBETT
210 CORBIN
211 CORCEGA
212 CORD
213 CORDELL
214 CORDES
215 CORDESTON
216 CORDOVA
217 CORDY
218 CORIFF
219 CORINTH
220 CORKSTONE
221 CORLENA
222 CORLETT
223 CORLEY
224 CORMANT
225 CORNELIA
226 CORNELIUS
227 CORNHILL
228 CORNICK
229 CORNING
230 CORNISH
231 CORNUTT
232 CORNVILLE
233 COROLLA
234 CORONA
235 CORONACA
236 COROZAL
237 COROZO
238 CORPENING
239 CORRAL
240 CORRALITOS
241 CORRECO
242 CORRIGAN
243 CORSON
244 CORTA
245 CORTADA
246 CORTEZ
247 CORTINA
248 CORUNNA
249 CORWIN
250 CORWITH
251 CORY
252 CORYDON
253 COSAD
254 COSEY
255 COSH
256 COSHOCTON
257 COSKI
258 COSTILLA
259 COSUMNES
260 COTACO
261 COTAIL
262 COTANT
263 COTATI
264 COTEAU
265 COTHA
266 COTITO
267 COTO
268 COTOPAXI
269 COTT
270 COTTER
271 COTTERAL
272 COTTONEVA
273 COTTONTHOMAS
274 COTTONWOOD
275 COTTRELL
276 COTULLA
277 COUCH
278 COUGARBAY
279 COUGHANOUR
280 COULSTONE

281 COULTERG
282 COULTERVILLE
283 COUNCELOR
284 COUNCIL
285 COUNTRYMAN
286 COUNTS
287 COUPEE
288 COUPEVILLE
289 COURT
290 COURTHOUSE
291 COURTLAND
292 COURTNEY
293 COURTROCK
294 COURVILLE
295 COUSE
296 COUSHATTA
297 COUTIS
298 COVE
299 COVELAND
300 COVELLO
301 COVERT
302 COVEYTOWN
303 COVILLE
304 COVING
305 COVINGTON
306 COWAN
307 COWARTS
308 COWCO
309 COWDEN
310 COWDREY
311 COWEEMAN
312 COWERS
313 COWETA
314 COWGIL
315 COWHORN
316 COWICHE
317 COWOOD
318 COWSLY
319 COWTON
320 COX
321 COXVILLE
322 COXWELL
323 COY
324 COYANOSA
325 COYATA
326 COYLE
327 COYNE
328 COZAD
329 COZBERG
330 COZTUR
331 CRABTREE
332 CRADDOCK
333 CRADLEBAUGH
334 CRAFT
335 CRAFTON
336 CRAGGEY
337 CRAGO
338 CRAGOLA
339 CRAGOSEN
340 CRAIG
341 CRAIGMILE
342 CRAIGSVILLE
343 CRAMER
344 CRAMONT
345 CRANE
346 CRANECREEK
347 CRANFILL
348 CRANNLER
349 CRANSTON
350 CRARY
351 CRASH
352 CRATER LAKE
353 CRATERMO
354 CRAVEN
355 CRAWFORD
356 CREAL
357 CREASEY
358 CREDO
359 CREED
360 CREEDMOOR
361 CREEMON
362 CREIGHTON
363 CRELDON
364 CREN
365 CRESBARD
366 CRESCO
367 CRESKEN
368 CRESPIN
369 CREST
370 CRESTLINE
371 CRESTMAN
372 CRESTVALE
373 CRETE
374 CREVA
375 CREVASSE
376 CREWS
377 CRIDER
378 CRIMS
379 CRINKER
380 CRIPPIN
381 CRISFIELD
382 CRISTO
383 CRISTOBAL
384 CRITCHELL
385 CRITTENDEN
386 CROATAN
387 CROCKER
388 CROCKETT
389 CROESUS
390 CROFTON
391 CROGHAN
392 CROKE
393 CROMWELL
394 CRONKHITE
395 CRONKS
396 CROOKED
397 CROOKED CREEK
398 CROOKSTON
399 CROOM
400 CROPLEY
401 CROQUIB
402 CROSBY
403 CROSIER
404 CROSS
405 CROSSPLAIN
406 CROSSTELL
407 CROSSVILLE
408 CROSWELL
409 CROT
410 CROTON
411 CROUCH
412 CROW
413 CROW CREEK
414 CROW FAMILY
415 CROW HILL
416 CROWCAMP
417 CROWFLATS
418 CROWFOOT
419 CROWHEART
420 CROWLEY
421 CROWNEST
422 CROWSHAW
423 CROWTHER
424 CROYDON
425 CROZIER
426 CRUCES
427 CRUCKTON
428 CRUICKSHANK
429 CRUISER
430 CRUMARINE
431 CRUME
432 CRUMP
433 CRUNKER
434 CRUTCH
435 CRUTCHER
436 CRYANDEPTS
437 CRYAQUENTS
438 CRYAQUEPTS
439 CRYAQUOLLS
440 CRYLUHA
441 CRYOBOROLLS
442 CRYOCHREPTS
443 CRYOFIBRISTS
444 CRYOFLUVENTS
445 CRYOHEMISTS
446 CRYOHUMODS
447 CRYORTHENTS
448 CRYORTHODS
449 CRYSTAL LAKE
450 CRYSTAL SPRINGS
451 CRYSTALBUTTE
452 CRYUMBREPTS
453 CUBA
454 CUBERANT
455 CUCAMUNGO
456 CUCHILLAS
457 CUDAHY
458 CUDDEBACK
459 CUERO
460 CUESTA
461 CUEVA
462 CUEVITAS
463 CUEVOLAND
464 CULBERTSON
465 CULDESAC
466 CULLABY
467 CULLEN
468 CULLEOKA
469 CULP
470 CULPEPER
471 CULTUS
472 CULVING
473 CUMBERLAND
474 CUMBRES
475 CUMLEY
476 CUMLIC HAPLOXERO
477 CUMMINGS
478 CUMULIC CRYOBOR
479 CUMULIC HAPLAQUO
480 CUMULIC HAPLOBOR
481 CUMULIC HAPLOXER
482 CUMULIC HAPLUDOL
483 CUMULIC HAPLUSTO
484 CUNARD
485 CUNDICK
486 CUNDIYO
487 CUNNINGHAM
488 CUPCO
489 CUPOLA
490 CUPPER
491 CUPPLES
492 CURABITH
493 CURANT
494 CURDLI
495 CURECANTI
496 CURHOLLOW
497 CURRAN
498 CURRIER
499 CURRITUCK
500 CURTIN
501 CURTIS CREEK
502 CURTIS SIDING
503 CURTISTOWN
504 CURTISVILLE
505 CUSHENBURY
506 CUSHING
507 CUSHMAN
508 CUSHOOL
509 CUSICK
510 CUSTCO
511 CUSTER
512 CUTAWAY
513 CUTHAND
514 CUTHBERT
515 CUTOFF
516 CUTSHIN
517 CUTZ
518 CUYAMA
519 CUYON
520 CYAN
521 CYCLONE
522 CYLINDER
523 CYMRIC
524 CYNTHIANA
525 CYNTHIANIA
526 CYPHER
527 CYRIL
528 CZAR
529 DABNEY
530 DABOB
531 DACKER
532 DACONO
533 DACORE
534 DACOSTA
535 DADE
536 DAGAN
537 DAGFLAT
538 DAGLUM
539 DAGOR
540 DAGUAO
541 DAGUEY
542 DAHLQUIST
543 DAICK
544 DAIGLE
545 DAILEY
546 DAINT
547 DAKENT
548 DAKOTA
549 DALBO
550 DALBY
551 DALCAN
552 DALCO
553 DALE
554 DALEVILLE
555 DALHART
556 DALIAN
557 DALIG
558 DALKENA
559 DALLAM
560 DALLARDSVILLE
561 DALLESPORT
562 DALTON
563 DALUPE
564 DALZELL
565 DAMASCUS
566 DAMEWOOD
567 DAMLUIS
568 DAMON
569 DANA
570 DANAHER
571 DANCY
572 DANDREA
573 DANDRIDGE
574 DANFORTH
575 DANGBERG
576 DANIA
577 DANJER
578 DANKO
579 DANLEY
580 DANN

581 DANNEMORA
582 DANSKIN
583 DANT
584 DANVERS
585 DANVILLE
586 DAPHNEDALE
587 DARBY
588 DARCO
589 DARDANELLE
590 DARDEN
591 DARDOOW
592 DARE
593 DARFUR
594 DARGOL
595 DARIEN
596 DARKBULL
597 DARL
598 DARLAND
599 DARLING
600 DARMSTADT
601 DARNELL
602 DARNEN
603 DAROW
604 DARR
605 DARRET
606 DARROCH
607 DARROUZETT
608 DARSIL
609 DARST
610 DART
611 DARVEY
612 DARWIN
613 DASHER
614 DASSEL
615 DAST
616 DATELAND
617 DATEMAN
618 DATIL
619 DATINO
620 DATWYLER
621 DAULTON
622 DAVEY
623 DAVIDELL
624 DAVIDSON
625 DAVIS
626 DAVISON
627 DAVTONE
628 DAWES
629 DAWHOO
630 DAWSON
631 DAWTONIA
632 DAXTY
633 DAY
634 DAYBELL
635 DAYSCHOOL
636 DAYTON
637 DAYTONA
638 DAYVILLE
639 DAZE
640 DE BACA
641 DE MASTERS
642 DE PERE
643 DEACON
644 DEADFALL
645 DEADHORSE
646 DEADMAN
647 DEADWOOD
648 DEAMA
649 DEAN
650 DEANDALE
651 DEARBORN
652 DEARYTON
653 DEATMAN
654 DEAVER
655 DEBAB
656 DEBACA
657 DEBENGER
658 DEBONE
659 DEBOOK
660 DEBORAH
661 DEBUTE
662 DECAN
663 DECANTEL
664 DECATHON
665 DECATUR
666 DECCA
667 DECHEL
668 DECKER
669 DECKERVILLE
670 DECLO
671 DECOLNEY
672 DECORDOVA
673 DECRAM
674 DECROSS
675 DECY
676 DEDAS
677 DEDMOUNT
678 DEE
679 DEECREE
680 DEEFAN
681 DEEMER
682 DEEPEEK
683 DEEPWATER
684 DEER CREEK
685 DEER PARK
686 DEERFIELD
687 DEERFORD
688 DEERHORN
689 DEERLODGE
690 DEERTON
691 DEERTRAIL
692 DEERWOOD
693 DEETZ
694 DEFIANCE
695 DEFORD
696 DEGARMO
697 DEGNER
698 DEGOLA
699 DEGRAND
700 DEGREY
701 DEHANA
702 DEHART
703 DEHAVEN
704 DEHLINGER
705 DEJARNET
706 DEKALB
707 DEKOOM
708 DEKOVEN
709 DEL REY
710 DELA
711 DELAMAR
712 DELANCO
713 DELAND
714 DELANEY
715 DELANO
716 DELASSUS
717 DELCOMB
718 DELDOTA
719 DELECO
720 DELENA
721 DELEON
722 DELETTE
723 DELFINA
724 DELFT
725 DELGADO
726 DELHI
727 DELICIAS
728 DELKS
729 DELL
730 DELLEKER
731 DELLO
732 DELLROSE
733 DELLS
734 DELMA
735 DELMITA
736 DELMONT
737 DELNORTE
738 DELORO
739 DELOSS
740 DELP
741 DELPHI
742 DELPHILL
743 DELPIEDRA
744 DELPLAIN
745 DELPOINT
746 DELRAY
747 DELRIDGE
748 DELSON
749 DELTON
750 DELWIN
751 DELYNDIA
752 DEMAR
753 DEMAST
754 DEMENT
755 DEMING
756 DEMKY
757 DEMNER
758 DEMONA
759 DEMONTREVILLE
760 DEMOPOLIS
761 DEMPSEY
762 DEMPSTER
763 DENAY
764 DENBY
765 DENCO
766 DENHAWKEN
767 DENMAN
768 DENMARK
769 DENNIS
770 DENNOT
771 DENNY
772 DENROCK
773 DENTON
774 DENURE
775 DENVER
776 DEODAR
777 DEPALT
778 DEPCOR
779 DEPORT
780 DEPUTY
781 DERA
782 DERALLO
783 DERB
784 DERBY
785 DERECHO
786 DERINDA
787 DERLY
788 DEROUX
789 DERRICK
790 DES MOINES
791 DESAN
792 DESART
793 DESATOYA
794 DESCALABRADO
795 DESCHELL
796 DESCHUTES
797 DESCOT
798 DESEED
799 DESERET
800 DESHA
801 DESHLER
802 DESKAMP
803 DESMET
804 DESPAIN
805 DESTAZO
806 DESTER
807 DETER
808 DETOUR
809 DETRA
810 DETRITAL
811 DETROIT
812 DEUNAH
813 DEV
814 DEVADA
815 DEVEN
816 DEVILS
817 DEVILSGAIT
818 DEVINE
819 DEVISADERO
820 DEVOE
821 DEVOIGNES
822 DEVOL
823 DEVORE
824 DEVOY
825 DEVRIES
826 DEWAR
827 DEWEY
828 DEWEYVILLE
829 DEWMINE
830 DEWVILLE
831 DEXTER
832 DIA
833 DIABLO
834 DIAMANTE
835 DIAMOND
836 DIAMOND SPRINGS
837 DIAMONDVILLE
838 DIANEV
839 DIANOLA
840 DIASPAR
841 DIATEE
842 DIAZ
843 DIBBLE
844 DIBOLL
845 DICK
846 DICKERSON
847 DICKEY
848 DICKINSON
849 DICKMAN
850 DICKSON
851 DIDDY
852 DIEHLSTADT
853 DIETRICH
854 DIGBY
855 DIGGER
856 DIGHTON
857 DIGIORGIO
858 DILANSON
859 DILL
860 DILLARD
861 DILLEY
862 DILLWYN
863 DILMAN
864 DILTON
865 DILTS
866 DIMAL
867 DIMMICK
868 DIMO
869 DIMYAW
870 DINA
871 DINCO
872 DINES
873 DINEVO
874 DINGLE
875 DINGLISHNA
876 DINKELMAN
877 DINKELS
878 DINNEN
879 DINSDALE
880 DINUBA

881 DINWOODY
882 DIOBSUD
883 DIOXICE
884 DIPMAN
885 DIPSEA
886 DIQUE
887 DIREGO
888 DISABEL
889 DISAUTEL
890 DISCO
891 DISHNER
892 DISHPAN
893 DISTERHEFF
894 DISTON
895 DITCHCAMP
896 DITHOD
897 DITNEY
898 DIVERS
899 DIVIDE
900 DIVOT
901 DIX
902 DIXALETA
903 DIXBORO
904 DIXIE
905 DIXMONT
906 DIXONVILLE
907 DIYOU
908 DOAK
909 DOBBINS
910 DOBBS
911 DOBEL
912 DOBENT
913 DOBROW
914 DOBY
915 DOCAS
916 DOCDEE
917 DOCENA
918 DOCKERY
919 DOCT
920 DODES
921 DODGE
922 DODGEVILLE
923 DODSON
924 DOEL
925 DOGER
926 DOGUE
927 DOLAND
928 DOLBEE
929 DOLEKEI
930 DOLEN
931 DOLES
932 DOLLAR
933 DOLLARD
934 DOLLARHIDE
935 DOLLYCLARK
936 DOLPH
937 DOLUS
938 DOME
939 DOMELL
940 DOMERIE
941 DOMEZ
942 DOMINGUEZ
943 DOMINIC
944 DOMINO
945 DOMINSON
946 DOMO
947 DONA ANA
948 DONAHUE
949 DONALD
950 DONALDSON
951 DONAVAN
952 DONERAIL
953 DONEY
954 DONICA
955 DONIPHAN
956 DONKEHILL
957 DONLONTON
958 DONNA
959 DONNAN
960 DONNARDO
961 DONNEL
962 DONNELLY
963 DONNER
964 DONNYBROOK
965 DOODLELINK
966 DOOLEY
967 DOOLIN
968 DOONE
969 DOOR
970 DORA
971 DORAN
972 DORB
973 DORCHESTER
974 DORERTON
975 DORMONT
976 DORNA
977 DOROSHIN
978 DOROTHEA
979 DOROVAN
980 DORRANCE
981 DORS
982 DORSET
983 DOSAMIGOS
984 DOSPALOS
985 DOSS
986 DOSSMAN
987 DOTARD
988 DOTEN
989 DOTHAN
990 DOTLAKE
991 DOTSERO
992 DOTTA
993 DOTY
994 DOUCETTE
995 DOUDLE
996 DOUDS
997 DOUGAL
998 DOUGAN
999 DOUGCLIFF
1000 DOUGH
1001 DOUGHERTY
1002 DOUGHTY
1003 DOUGLAS
1004 DOUGVILLE
1005 DOUHIDE
1006 DOURO
1007 DOVER
1008 DOVRAY
1009 DOW
1010 DOWAGIAC
1011 DOWDE
1012 DOWELLTON
1013 DOWNATA
1014 DOWNER
1015 DOWNEY
1016 DOWNEYVILLE
1017 DOWNS
1018 DOYCE
1019 DOYLESTOWN
1020 DOYN
1021 DRA
1022 DRAGE
1023 DRAGOON
1024 DRAGSTON
1025 DRAKE
1026 DRAKNAB
1027 DRALL
1028 DRANYON
1029 DRAPER
1030 DRAX
1031 DREDGE
1032 DRESDEN
1033 DRESSLER
1034 DREWING
1035 DREWS
1036 DREXEL
1037 DRIFTWOOD
1038 DRIGGS
1039 DRISCOLL
1040 DRIT
1041 DRIVER
1042 DROEM
1043 DRUM
1044 DRUMMER
1045 DRUMMOND
1046 DRURY
1047 DRY CREEK
1048 DRYADINE
1049 DRYBURG
1050 DRYDEN
1051 DRYN
1052 DU PAGE
1053 DUANE
1054 DUART
1055 DUBAKELLA
1056 DUBAY
1057 DUBBS
1058 DUBLON
1059 DUBOIS
1060 DUBUQUE
1061 DUCHESNE
1062 DUCKHILL
1063 DUCKREE
1064 DUCKSTON
1065 DUCO
1066 DUDA
1067 DUDLEY
1068 DUEL
1069 DUELM
1070 DUETTE
1071 DUFF
1072 DUFFAU
1073 DUFFER
1074 DUFFERN
1075 DUFFIELD
1076 DUFFSON
1077 DUFORT
1078 DUFUR
1079 DUGGINS
1080 DUGOUT
1081 DUGWAY
1082 DUKES
1083 DULAC
1084 DULCE
1085 DULLES
1086 DULUTH
1087 DUMAS
1088 DUMFRIES
1089 DUMMERSTON
1090 DUMONT
1091 DUMPS
1092 DUN GLEN
1093 DUNBAR
1094 DUNBARTON
1095 DUNBRIDGE
1096 DUNC
1097 DUNCAN
1098 DUNCANNON
1099 DUNCKLEY
1100 DUNCOM
1101 DUNDAS
1102 DUNDAY
1103 DUNDEE
1104 DUNE LAND
1105 DUNELLEN
1106 DUNFORD
1107 DUNGENESS
1108 DUNKIRK
1109 DUNLATOP
1110 DUNMORE
1111 DUNN
1112 DUNNIGAN
1113 DUNNING
1114 DUNNLAKE
1115 DUNNVILLE
1116 DUNOIR
1117 DUNPHY
1118 DUNSMUIR
1119 DUNTON
1120 DUNUL
1121 DUPEE
1122 DUPLIN
1123 DUPO
1124 DUPONT
1125 DUPREE
1126 DURADOS
1127 DURALDE
1128 DURAND
1129 DURANGO
1130 DURANT
1131 DURARGIDIC ARENT
1132 DURBIN
1133 DURELLE
1134 DURFEE
1135 DURHAM
1136 DURKEE
1137 DUROC
1138 DURRSTEIN
1139 DURST
1140 DUSLER
1141 DUSTON
1142 DUTCHESS
1143 DUTEK
1144 DUTTON
1145 DUVAL
1146 DUXBURY
1147 DUZEL
1148 DWIGHT
1149 DWORSHAK
1150 DWYER
1151 DYE
1152 DYKE
1153 DYLAN
1154 DYRENG
1155 DYSTRANDEPTS
1156 DYSTRIC XEROCHRE
1157 DYSTRIC XERORTHE
1158 DYSTROCHREPTIC A
1159 EACHUS
1160 EACHUSTON
1161 EAGAR
1162 EAGLECONE
1163 EAGLEPASS
1164 EAGLEVILLE
1165 EAKIN
1166 EALY
1167 EAPA
1168 EARCREE
1169 EARLE
1170 EARLMONT
1171 EARP
1172 EARSMAN
1173 EASLEY
1174 EASPUR
1175 EAST FORK
1176 EAST LAKE
1177 EASTABLE
1178 EASTCAN
1179 EASTCHOP
1180 EASTGATE

1181 EASTLAND
1182 EASTON
1183 EASTONVILLE
1184 EASTPORT
1185 EASTWELL
1186 EATON
1187 EAUGALLIE
1188 EAUPLEINE
1189 EBA
1190 EBAL
1191 EBBERT
1192 EBBS
1193 EBERLY
1194 EBIC
1195 EBODA
1196 EBON
1197 EBRO
1198 ECCLES
1199 ECHARD
1200 ECHAW
1201 ECHEMOOR
1202 ECKERT
1203 ECKLEY
1204 ECKMAN
1205 ECKRANT
1206 ECKVOLL
1207 ECOLA
1208 ECONFINA
1209 ECTOR
1210 EDALGO
1211 EDDINGS
1212 EDDS
1213 EDDY
1214 EDEN
1215 EDENBOWER
1216 EDENTON
1217 EDFRO
1218 EDGAR
1219 EDGEHILL
1220 EDGELEY
1221 EDGEMONT
1222 EDGEWATER
1223 EDGEWICK
1224 EDGINGTON
1225 EDINA
1226 EDINBURG
1227 EDISTO
1228 EDLOE
1229 EDMINSTER
1230 EDMONDS
1231 EDMORE
1232 EDMUND
1233 EDMUNDSTON
1234 EDNA
1235 EDNEYTOWN
1236 EDNEYVILLE
1237 EDOM
1238 EDROY
1239 EDSON
1240 EDWARDS
1241 EEL
1242 EELCOVE
1243 EELPOINT
1244 EEP
1245 EFFIE
1246 EFFINGTON
1247 EGAM
1248 EGAN
1249 EGAS
1250 EGBERT
1251 EGELAND
1252 EGINBENCH
1253 EGLIN
1254 EGYPT
1255 EICKS
1256 EIGHTLAR
1257 EIGHTMILE
1258 EILERTSEN
1259 EITZEN
1260 EKAH
1261 EKALAKA
1262 EKRUB
1263 EL DARA
1264 EL PECO
1265 EL RANCHO
1266 EL SOLYO
1267 ELANDCO
1268 ELBA
1269 ELBAVILLE
1270 ELBERT
1271 ELBETH
1272 ELBON
1273 ELBURN
1274 ELBUTTE
1275 ELCO
1276 ELD
1277 ELDEAN
1278 ELDER
1279 ELDER HOLLOW
1280 ELDERON
1281 ELDGIN
1282 ELDON
1283 ELDORADO
1284 ELDRIDGE
1285 ELECTRA
1286 ELEROY
1287 ELEVA
1288 ELFRIDA
1289 ELGEE
1290 ELIJAH
1291 ELINDIO
1292 ELIOAK
1293 ELIZA
1294 ELK
1295 ELK HOLLOW
1296 ELK MOUNTAIN
1297 ELKADER
1298 ELKCREEK
1299 ELKHART
1300 ELKHILLS
1301 ELKHORN
1302 ELKINS
1303 ELKINSVILLE
1304 ELKMOUND
1305 ELKNER
1306 ELKOL
1307 ELKSEL
1308 ELKTON
1309 ELLABELLE
1310 ELLEDGE
1311 ELLEN
1312 ELLETT
1313 ELLIBER
1314 ELLICOTT
1315 ELLINGTON
1316 ELLINOR
1317 ELLIOTT
1318 ELLIOTTSVILLE
1319 ELLIS
1320 ELLISFORDE
1321 ELLISVILLE
1322 ELLOAM
1323 ELLOREE
1324 ELLSWORTH
1325 ELLUM
1326 ELLZEY
1327 ELM LAKE
1328 ELMDALE
1329 ELMENDORF
1330 ELMINA
1331 ELMIRA
1332 ELMO
1333 ELMONT
1334 ELMORE
1335 ELMRIDGE
1336 ELMVILLE
1337 ELMWOOD
1338 ELNIDO
1339 ELNORA
1340 ELO
1341 ELOCHOMAN
1342 ELOIKA
1343 ELOMA
1344 ELPAM
1345 ELPEDRO
1346 ELRED
1347 ELRICK
1348 ELRIN
1349 ELROD
1350 ELROSE
1351 ELS
1352 ELSAH
1353 ELSINBORO
1354 ELSMERE
1355 ELSTON
1356 ELTREE
1357 ELTSAC
1358 ELVADA
1359 ELVE
1360 ELVEDERE
1361 ELVERS
1362 ELVIRA
1363 ELWELL
1364 ELWHA
1365 ELWOOD
1366 ELY
1367 ELYSIAN
1368 ELZINGA
1369 EMBAL
1370 EMBARGO
1371 EMBDEN
1372 EMBERTON
1373 EMBLEM
1374 EMBRY
1375 EMBUDO
1376 EMDENT
1377 EMERALD
1378 EMERALDA
1379 EMERSON
1380 EMIGRANT
1381 EMIGRATION
1382 EMILY
1383 EMMA
1384 EMMERT
1385 EMMET
1386 EMORY
1387 EMOT
1388 EMPEDRADO
1389 EMPEYVILLE
1390 EMPIRE
1391 EMPORIA
1392 EMRICK
1393 EMRO
1394 ENBAR
1395 ENCAMPMENT
1396 ENCHANTED
1397 ENCIERRO
1398 ENCINA
1399 ENDCAV
1400 ENDERS
1401 ENDERSBY
1402 ENDICOTT
1403 ENDLICH
1404 ENDSAW
1405 ENERGY
1406 ENET
1407 ENFIELD
1408 ENGELHARD
1409 ENGETT
1410 ENGLE
1411 ENGLEWOOD
1412 ENKO
1413 ENLOE
1414 ENNING
1415 ENNIS
1416 ENOCH
1417 ENOCHVILLE
1418 ENON
1419 ENOREE
1420 ENOS
1421 ENOSBURG
1422 ENSENADA
1423 ENSIGN
1424 ENSLEY
1425 ENSTROM
1426 ENTENTE
1427 ENTERO
1428 ENTERPRISE
1429 ENTIAT
1430 ENTMOOT
1431 ENZIAN
1432 EOJ
1433 EPHRAIM
1434 EPHRATA
1435 EPIKOM
1436 EPLEY
1437 EPOKE
1438 EPOUFETTE
1439 EPPING
1440 EPSIE
1441 ERA
1442 ERAKATAK
1443 ERAM
1444 ERAMOSH
1445 ERBER
1446 ERCAN
1447 ERD
1448 ERICSON
1449 ERIE
1450 ERIN
1451 ERNEM
1452 ERNEST
1453 ERNO
1454 ERRAMOUSPE
1455 ERVIDE
1456 ESCABOSA
1457 ESCALANTE
1458 ESCAMBIA
1459 ESCANO
1460 ESCARLO
1461 ESCONDIDO
1462 ESHAMY
1463 ESMERALDA
1464 ESMOND
1465 ESPARTO
1466 ESPELIE
1467 ESPIL
1468 ESPINA
1469 ESPINAL
1470 ESPINOSA
1471 ESPLIN
1472 ESPY
1473 ESQUATZEL
1474 ESRO
1475 ESS
1476 ESSAL
1477 ESSEN
1478 ESSEX
1479 ESSEXVILLE
1480 ESTACADO

1481 ESTACION
1482 ESTATE
1483 ESTELLINE
1484 ESTER
1485 ESTERO
1486 ESTHERVILLE
1487 ESTIVE
1488 ESTO
1489 ESTRELLA
1490 ETACH
1491 ETCHEN
1492 ETELKA
1493 ETHAN
1494 ETHANIA
1495 ETHELMAN
1496 ETHETE
1497 ETHRIDGE
1498 ETIL
1499 ETOE
1500 ETOILE
1501 ETOWAH
1502 ETOWN
1503 ETSEL
1504 ETTA
1505 ETTER
1506 ETTRICK
1507 EUBANKS
1508 EUCLID
1509 EUDORA
1510 EUER
1511 EUFAULA
1512 EUHARLEE
1513 EULONIA
1514 EUNOLA
1515 EUREKA
1516 EUSBIO
1517 EUSTIS
1518 EUTAW
1519 EVADALE
1520 EVANGELINE
1521 EVANS
1522 EVANSTON
1523 EVANSVILLE
1524 EVANT
1525 EVARD
1526 EVARO
1527 EVART
1528 EVENDALE
1529 EVERETT
1530 EVERGLADES
1531 EVERLY
1532 EVERMAN
1533 EVERSON
1534 EVERWHITE
1535 EVESBORO
1536 EVRIDGE
1537 EWA
1538 EWALL
1539 EXCELSIOR
1540 EXCHEQUER
1541 EXEL
1542 EXETER
1543 EXETTE
1544 EXIRA
1545 EXLINE
1546 EXPRESS
1547 EXRAY
1548 EXUM
1549 EYAK
1550 EYERBOW
1551 EYLAU
1552 EYOTA
1553 EYRE
1554 FABIUS
1555 FACEVILLE
1556 FACEY
1557 FACTORY
1558 FADDIN
1559 FAGAN
1560 FAGASA
1561 FAHEY
1562 FAIM
1563 FAIRBANKS
1564 FAIRCHILD
1565 FAIRDALE
1566 FAIRFAX
1567 FAIRFIELD
1568 FAIRHAVEN
1569 FAIRLIE
1570 FAIRLO
1571 FAIRMOUNT
1572 FAIRPLAY
1573 FAIRPOINT
1574 FAIRPORT
1575 FAIRWAY
1576 FAIRYDELL
1577 FAJARDO
1578 FALAYA
1579 FALBA
1580 FALCON
1581 FALFA
1582 FALFURRIAS
1583 FALK
1584 FALKIRK
1585 FALKNER
1586 FALLBROOK
1587 FALLCREEK
1588 FALLERT
1589 FALLON
1590 FALLSAM
1591 FALLSINGTON
1592 FALOMA
1593 FALULA
1594 FANAL
1595 FANDANGLE
1596 FANDOW
1597 FANG
1598 FANNIN
1599 FANNO
1600 FANTZ
1601 FANU
1602 FAPS
1603 FARAWAY
1604 FARB
1605 FARBER
1606 FARGO
1607 FARISITA
1608 FARLAND
1609 FARLOW
1610 FARMINGTON
1611 FARMSWORTH
1612 FARMTON
1613 FARNHAM
1614 FARNHAMTON
1615 FARNUF
1616 FARNUM
1617 FARRAGUT
1618 FARRAR
1619 FARRELL
1620 FARRENBURG
1621 FARROT
1622 FARVA
1623 FASHING
1624 FASKIN
1625 FATHOM
1626 FATIMA
1627 FATTIG
1628 FAUQUIER
1629 FAUSSE
1630 FAVRET
1631 FAWIN
1632 FAX
1633 FAXON
1634 FAYETTE
1635 FAYETTEVILLE
1636 FAYWOOD
1637 FE
1638 FEATHERLEGS
1639 FEDJI
1640 FEDORA
1641 FELAN
1642 FELDA
1643 FELICITY
1644 FELIPE
1645 FELKER
1646 FELLOWSHIP
1647 FELOR
1648 FELT
1649 FELTA
1650 FELTHAM
1651 FELTNER
1652 FELTON
1653 FELTONIA
1654 FENCE
1655 FENDALL
1656 FENN
1657 FENSTER
1658 FENWOOD
1659 FERA
1660 FERDELFORD
1661 FERDINAND
1662 FERGUS
1663 FERN CLIFF
1664 FERNANDO
1665 FERNHAVEN
1666 FERNLEY
1667 FERNOW
1668 FERNPOINT
1669 FERNWOOD
1670 FERRELO
1671 FERRIS
1672 FERROBURRO
1673 FERRON
1674 FERTALINE
1675 FERTEG
1676 FESTINA
1677 FETT
1678 FETTIC
1679 FETZER
1680 FIANDER
1681 FIDALGO
1682 FIDDLER
1683 FIDDLETOWN
1684 FIDDYMENT
1685 FIELD
1686 FIELDING
1687 FIELDON
1688 FIFER
1689 FIFIELD
1690 FILION
1691 FILIRAN
1692 FILLMORE
1693 FINCASTLE
1694 FINCH
1695 FINCHFORD
1696 FINDOUT
1697 FINGAL
1698 FINGEROCK
1699 FINLAND
1700 FINLEY
1701 FINLEYPOINT
1702 FINNERTY
1703 FINOL
1704 FINROD
1705 FIONE
1706 FIRADA
1707 FIREBALL
1708 FIREBOX
1709 FIRESTEEL
1710 FIRESTONE
1711 FIRMAGE
1712 FIRO
1713 FIROKE
1714 FIRSTVIEW
1715 FIRTH
1716 FISHERS
1717 FISHHOOK
1718 FISHLAKE
1719 FISHPOT
1720 FISHTRAP
1721 FISK
1722 FITCHVILLE
1723 FITZGERALD
1724 FITZHUGH
1725 FIVEMILE
1726 FIVEOH
1727 FIVES
1728 FLAGG
1729 FLAGLER
1730 FLAGSTAFF
1731 FLAK
1732 FLAMING
1733 FLANAGAN
1734 FLANDREAU
1735 FLANE
1736 FLASHER
1737 FLATHEAD
1738 FLATHORN
1739 FLATIRONS
1740 FLATNOSE
1741 FLATRON
1742 FLATTOP
1743 FLAXTON
1744 FLEAK
1745 FLEER
1746 FLEISCHMANN
1747 FLEMING
1748 FLEMINGTON
1749 FLETCHER
1750 FLEX
1751 FLOER
1752 FLOKE
1753 FLOM
1754 FLOMATON
1755 FLOMOT
1756 FLORALA
1757 FLORENCE
1758 FLORESVILLE
1759 FLORIDANA
1760 FLORIN
1761 FLORISSANT
1762 FLORITA
1763 FLOTAG
1764 FLOWELL
1765 FLOWEREE
1766 FLOYD
1767 FLUETSCH
1768 FLUGLE
1769 FLUVANNA
1770 FLUVAQUENTIC HAP
1771 FLUVAQUENTIC HAP
1772 FLUVAQUENTIC HU
1773 FLUVAQUENTS
1774 FLUVENTIC HAPLOB
1775 FLUVENTIC HAPLOX
1776 FLUVENTIC HAPLUS
1777 FLYBOW
1778 FLYGARE
1779 FOARD
1780 FOEHLIN

1781 FOLA
1782 FOLDAHL
1783 FOLEY
1784 FOLLET
1785 FOMSENG
1786 FONDA
1787 FONDIS
1788 FONNER
1789 FONS
1790 FONTANA
1791 FONTREEN
1792 FOPIANO
1793 FORADA
1794 FORAKER
1795 FORBES
1796 FORBING
1797 FORBSVILLE
1798 FORD
1799 FORDICE
1800 FORDNEY
1801 FORDTRAN
1802 FORDUM
1803 FORDVILLE
1804 FORELAND
1805 FORELLE
1806 FORESMAN
1807 FORESTBURG
1808 FORESTDALE
1809 FORESTER
1810 FORESTON
1811 FORGAY
1812 FORK
1813 FORKWOOD
1814 FORMADER
1815 FORMAN
1816 FORMDALE
1817 FORNEY
1818 FORNOR
1819 FORREST
1820 FORSEER
1821 FORSEY
1822 FORSGREN
1823 FORSYTH
1824 FORT COLLINS
1825 FORT MEADE
1826 FORT MEADE VARIA
1827 FORT MOTT
1828 FORT ROCK
1829 FORTANK
1830 FORTESCUE
1831 FORTUNA
1832 FORTWINGATE
1833 FORVIC
1834 FORWARD
1835 FOSS
1836 FOSSILON
1837 FOSSUM
1838 FOSTER
1839 FOSTORIA
1840 FOUNTAIN
1841 FOUR STAR
1842 FOURCHE
1843 FOURLOG
1844 FOURMILE
1845 FOX
1846 FOXCREEK
1847 FOXHOME
1848 FOXMOUNT
1849 FOXOL
1850 FOXTON
1851 FOXWORTH
1852 FRADDLE
1853 FRAGIUDALFS, SIL
1854 FRAILEY
1855 FRAILTON
1856 FRAM
1857 FRANCIS
1858 FRANCISCAN
1859 FRANCITAS
1860 FRANDSEN
1861 FRANKFORT
1862 FRANKIRK
1863 FRANKLIN
1864 FRANKSTOWN
1865 FRANKTOWN
1866 FRANKVILLE
1867 FRATERNIDAD
1868 FRAVAL
1869 FRAZER
1870 FRAZERTON
1871 FRED
1872 FREDENSBORG
1873 FREDERICK
1874 FREDON
1875 FREDONIA
1876 FREDONYER
1877 FREEBURG
1878 FREECE
1879 FREEDOM
1880 FREEHOLD
1881 FREEMAN
1882 FREEMANVILLE
1883 FREEON
1884 FREER
1885 FREEST
1886 FREESTONE
1887 FREETOWN
1888 FREEWATER
1889 FREEZENER
1890 FREEZEOUT
1891 FRELSBURG
1892 FREMONT
1893 FREN
1894 FRENCH
1895 FRENCHCREEK
1896 FRENCHMAN
1897 FRENCHTOWN
1898 FRESHWATER
1899 FRESNO
1900 FREWA
1901 FREZNIK
1902 FRIANA
1903 FRIANT
1904 FRIDLO
1905 FRIENDS
1906 FRIENDSHIP
1907 FRIES
1908 FRIESLAND
1909 FRIJOLES
1910 FRINDLE
1911 FRIO
1912 FRIONA
1913 FRIOTON
1914 FRIPP
1915 FRISCO
1916 FRIZZELL
1917 FROBERG
1918 FRODO
1919 FROHMAN
1920 FROLIC
1921 FRONDORF
1922 FRONTENAC
1923 FRONTON
1924 FROST
1925 FROZARD
1926 FRUITA
1927 FRUITFIELD
1928 FRUITHURST
1929 FRUITLAND
1930 FRYE
1931 FRYEBURG
1932 FT. DRUM
1933 FT. GREEN
1934 FUBBLE
1935 FUEGO
1936 FUERA
1937 FUGAWEE
1938 FUGHES
1939 FULCHER
1940 FULDA
1941 FULLAM
1942 FULLER
1943 FULLERTON
1944 FULMER
1945 FULSHEAR
1946 FULSTONE
1947 FULTON
1948 FULTS
1949 FULWIDER
1950 FUNTER
1951 FUQUAY
1952 FURNISS
1953 FURY
1954 FUSULINA
1955 GAASTRA
1956 GABALDON
1957 GABBS
1958 GABBVALLY
1959 GABEL
1960 GABICA
1961 GABINO
1962 GACEY
1963 GACHADO
1964 GACIBA
1965 GADDES
1966 GADDY
1967 GADSDEN
1968 GAGEBY
1969 GAGETOWN
1970 GAGIL
1971 GAHEE
1972 GAIB
1973 GAINES
1974 GAINESBORO
1975 GAINESVILLE
1976 GALATA
1977 GALBRETH
1978 GALCHUTT
1979 GALE
1980 GALEN
1981 GALEPPI
1982 GALESTOWN
1983 GALEY
1984 GALFAND
1985 GALILEE
1986 GALISTEO
1987 GALLAND
1988 GALLATIN
1989 GALLEGOS
1990 GALLEN
1991 GALLEPI
1992 GALLIA
1993 GALLIME
1994 GALLION
1995 GALLMAN
1996 GALLUP
1997 GALOO
1998 GALT
1999 GALVA
2000 GALVESTON
2001 GALVEZ
2002 GALVIN
2003 GALWAY
2004 GAMBLER
2005 GAMBOA
2006 GAMGEE
2007 GANCE
2008 GANDO
2009 GANIS
2010 GANNETT
2011 GANSNER
2012 GAPCOT
2013 GAPO
2014 GAPPMAYER
2015 GARA
2016 GARBER
2017 GARBO
2018 GARBUTT
2019 GARCENO
2020 GARCES
2021 GARCIA
2022 GARCITAS
2023 GARCON
2024 GARDELLA
2025 GARDENA
2026 GARDINER
2027 GARDNER'S FORK
2028 GARDNERVILLE
2029 GARDONE
2030 GAREY
2031 GARFAN
2032 GARFIELD
2033 GARIPER
2034 GARITA
2035 GARLAND
2036 GARLET
2037 GARLOCK
2038 GARMON
2039 GARMORE
2040 GARNEL
2041 GARNER
2042 GARNES
2043 GARO
2044 GARR
2045 GARRETSON
2046 GARRETT
2047 GARRISON
2048 GARROCHALES
2049 GARSID
2050 GARTON
2051 GARVESON
2052 GARVIN
2053 GARWIN
2054 GARZA
2055 GAS CREEK
2056 GASCONADE
2057 GASIL
2058 GASQUET
2059 GASSVILLE
2060 GATES
2061 GATESON
2062 GATEVIEW
2063 GATEWAY
2064 GATEWOOD
2065 GATLIN
2066 GATOR
2067 GATTON
2068 GAULDY
2069 GAVILAN
2070 GAVINS
2071 GAVIOTA
2072 GAY
2073 GAYLESVILLE
2074 GAYLORD
2075 GAYNOR
2076 GAYVILLE
2077 GAZELLE
2078 GAZOS
2079 GEARHART
2080 GEARY

2081 GED
2082 GEE
2083 GEEBURG
2084 GEEMORE
2085 GEER
2086 GEERTSEN
2087 GEFO
2088 GEISEL
2089 GELKIE
2090 GELKIE, MODERATE
2091 GEM
2092 GEMID
2093 GEMSON
2094 GENAW
2095 GENEGRAF
2096 GENESEE
2097 GENEVA
2098 GENOA
2099 GENOLA
2100 GENTILLY
2101 GENTRY
2102 GEOCONDA
2103 GEOHROCK
2104 GEORGETOWN
2105 GEORGEVILLE
2106 GEORGIA
2107 GEPFORD
2108 GEPP
2109 GEPPERT
2110 GERALD
2111 GERBER
2112 GERDRUM
2113 GERING
2114 GERLACH
2115 GERLANE
2116 GERLE
2117 GERMANTOWN
2118 GERMANY
2119 GERRARD
2120 GERST
2121 GESSIE
2122 GESSNER
2123 GESTRIN
2124 GETAWAY
2125 GETCHELL
2126 GETTYS
2127 GETZVILLE
2128 GEYSEN
2129 GIBBLER
2130 GIBBON
2131 GIDEON
2132 GIFFORD
2133 GIGGER
2134 GILA
2135 GILBERT
2136 GILBY
2137 GILCHRIST
2138 GILCO
2139 GILCREST
2140 GILEAD
2141 GILES
2142 GILFORD
2143 GILISPIE
2144 GILLAND
2145 GILLENDER
2146 GILLIAM
2147 GILLIGAN
2148 GILLS
2149 GILLSBURG
2150 GILMAN
2151 GILMORE
2152 GILPAR
2153 GILPIN
2154 GILROY
2155 GILSTON
2156 GILT EDGE
2157 GINAT
2158 GINGER
2159 GINI
2160 GINLAND
2161 GINNIS
2162 GINSER
2163 GIRARD
2164 GIRARDOT
2165 GIRD
2166 GIST
2167 GITAKUP
2168 GITAM
2169 GIVIN
2170 GLACIERCREEK
2171 GLACIERS
2172 GLADDEN
2173 GLADEL
2174 GLADEVILLE
2175 GLADEWATER
2176 GLADWIN
2177 GLANN
2178 GLASGOW
2179 GLASSNER
2180 GLEAN
2181 GLEASON
2182 GLEBE
2183 GLEN
2184 GLENBAR
2185 GLENBERG
2186 GLENBROOK
2187 GLENCARB
2188 GLENCOE
2189 GLENDALE
2190 GLENDERSON
2191 GLENDIVE
2192 GLENDORA
2193 GLENEDEN
2194 GLENELG
2195 GLENFORD
2196 GLENGARY
2197 GLENHALL
2198 GLENHAM
2199 GLENMAN
2200 GLENMORA
2201 GLENNALLEN
2202 GLENOMA
2203 GLENPOOL
2204 GLENRIO
2205 GLENROSE
2206 GLENROSS
2207 GLENSTED
2208 GLENTON
2209 GLENTOSH
2210 GLENVIEW
2211 GLENVILLE
2212 GLOHM
2213 GLORIA
2214 GLOUCESTER
2215 GLOVER
2216 GLYNDON
2217 GLYNN
2218 GLYNWOOD
2219 GLYPHS
2220 GOBAR
2221 GOBERNADOR
2222 GOBINE
2223 GOBLE
2224 GOBLIN
2225 GOCHEA
2226 GODDARD
2227 GODDE
2228 GODECKE
2229 GODFREY
2230 GODWIN
2231 GOEMMER
2232 GOESLING
2233 GOESSEL
2234 GOGEBIC
2235 GOL
2236 GOLCONDA
2237 GOLD CREEK
2238 GOLDBERG
2239 GOLDENDALE
2240 GOLDFINCH
2241 GOLDHILL
2242 GOLDMAN
2243 GOLDMIRE
2244 GOLDRIDGE
2245 GOLDRUN
2246 GOLDSBORO
2247 GOLDSTON
2248 GOLDSTREAM
2249 GOLDUST
2250 GOLDVALE
2251 GOLDVEIN
2252 GOLDYKE
2253 GOLETA
2254 GOLIAD
2255 GOLSUM
2256 GOLTRY
2257 GOLVA
2258 GOMERY
2259 GOMEZ
2260 GONVICK
2261 GONZAGA
2262 GOOCH
2263 GOODING
2264 GOODINGTON
2265 GOODLAND
2266 GOODLOW
2267 GOODMAN
2268 GOODNIGHT
2269 GOODPASTER
2270 GOODRICH
2271 GOODSPRINGS
2272 GOODWILL
2273 GOODWIN
2274 GOOSE CREEK
2275 GOOSE LAKE
2276 GOOSMUS
2277 GORDO
2278 GORE
2279 GOREEN
2280 GORGAS
2281 GORGONIO
2282 GORHAM
2283 GORIN
2284 GORING
2285 GORMAN
2286 GORSKEL
2287 GORSOR
2288 GORST
2289 GORUS
2290 GORZELL
2291 GOSA
2292 GOSHEN
2293 GOSHUTE
2294 GOSINTA
2295 GOSLIN
2296 GOSNEY
2297 GOSPER
2298 GOSPORT
2299 GOSS
2300 GOSUMI
2301 GOTEBO
2302 GOTHAM
2303 GOTHARD
2304 GOTHENBURG
2305 GOTHIC
2306 GOTHO
2307 GOULDING
2308 GOURLEY
2309 GOVE
2310 GOWEN
2311 GOWKER
2312 GOWTON
2313 GOZEM
2314 GOZEN
2315 GRABE
2316 GRABLE
2317 GRACEMONT
2318 GRACEMORE
2319 GRACEVILLE
2320 GRADON
2321 GRADY
2322 GRAFEN
2323 GRAFTON
2324 GRAGS
2325 GRAHAM
2326 GRAIL
2327 GRAINOLA
2328 GRALEY
2329 GRALIC
2330 GRAN
2331 GRANATH
2332 GRANBY
2333 GRANDE RONDE
2334 GRANDFIELD
2335 GRANDMORE
2336 GRANDPON
2337 GRANDVIEW
2338 GRANER
2339 GRANGEMONT
2340 GRANGEVILLE
2341 GRANILE
2342 GRANO
2343 GRANT
2344 GRANTFORK
2345 GRANTHAM
2346 GRANTSBURG
2347 GRANTSDALE
2348 GRANVILLE
2349 GRANYON
2350 GRANZAN
2351 GRAPEVINE
2352 GRASMERE
2353 GRASSNA
2354 GRASSVAL
2355 GRASSVALLEY
2356 GRASSY BUTTE
2357 GRASSYCONE
2358 GRAT
2359 GRATTAN
2360 GRAUFELS
2361 GRAVDEN
2362 GRAVEL PIT
2363 GRAVELLY LAND
2364 GRAVELTON
2365 GRAYBERT
2366 GRAYCALM
2367 GRAYFORD
2368 GRAYLAND
2369 GRAYLING
2370 GRAYLOCK
2371 GRAYPOINT
2372 GRAYROCK
2373 GRAYS
2374 GREAT BEND
2375 GREDGE
2376 GREEN BLUFF
2377 GREEN CANYON
2378 GREEN RIVER
2379 GREENBRAE
2380 GREENCREEK

2381 GREENDALE
2382 GREENE
2383 GREENFIELD
2384 GREENHALGH
2385 GREENHORN
2386 GREENLEAF
2387 GREENMAN
2388 GREENOUGH
2389 GREENSON
2390 GREENTON
2391 GREENVILLE
2392 GREENVINE
2393 GREENWATER
2394 GREENWAY
2395 GREENWOOD
2396 GREHALEM
2397 GRELL
2398 GRELLTON
2399 GRENADA
2400 GRENADIER
2401 GRENVILLE
2402 GRESHAM
2403 GRETDIVID
2404 GREWINGK
2405 GREYBACK
2406 GREYBO
2407 GREYBULL
2408 GREYEAGLE
2409 GREYPOINT
2410 GREYS
2411 GRIBBLE
2412 GRIDELL
2413 GRIDGE
2414 GRIDLEY
2415 GRIETA
2416 GRIEVES
2417 GRIFFITH
2418 GRIFFY
2419 GRIFTON
2420 GRIGSBY
2421 GRIGSTON
2422 GRIMM
2423 GRIMSLEY
2424 GRIMSTAD
2425 GRIMSTONE
2426 GRINA
2427 GRIND BROOK
2428 GRINDBROOK
2429 GRINDSTONE
2430 GRINK
2431 GRINROD
2432 GRISDALE
2433 GRISWOLD
2434 GRITNEY
2435 GRIVER
2436 GRIZZLY
2437 GROBUTTE
2438 GROGAN
2439 GROOM
2440 GROSECLOSE
2441 GROSS
2442 GROTON
2443 GROTTE
2444 GROTTO
2445 GROUSECREEK
2446 GROUSEVILLE
2447 GROVE
2448 GROVECITY
2449 GROVENA
2450 GROVER
2451 GROVETON
2452 GROWDEN
2453 GROWLER
2454 GROWTON
2455 GRUBBS
2456 GRUBSTAKE
2457 GRUENE
2458 GRULLA
2459 GRUMMIT
2460 GRUNDY
2461 GRUVER
2462 GRYGLA
2463 GSCHWEND
2464 GUADALUPE
2465 GUAJE
2466 GUAMANI
2467 GUANABANO
2468 GUANAJIBO
2469 GUANICA
2470 GUARD
2471 GUAYABO
2472 GUAYABOTA
2473 GUAYAMA
2474 GUBE
2475 GUBEN
2476 GUCKEEN
2477 GUDGREY
2478 GUELPH
2479 GUEMES
2480 GUENOC
2481 GUENTHER
2482 GUERNSEY
2483 GUERRERO
2484 GUEST
2485 GUFFIN
2486 GUGUAK
2487 GUILDER
2488 GUISER
2489 GULER
2490 GULF
2491 GULLIED LAND
2492 GUMBLE
2493 GUMBOOT
2494 GUNBARREL
2495 GUND
2496 GUNDY
2497 GUNLOCK
2498 GUNN
2499 GUNNEL
2500 GUNSIGHT
2501 GUNSONE
2502 GUNTER
2503 GUP
2504 GURDANE
2505 GURDON
2506 GURLEY
2507 GURNEY
2508 GUSTIN
2509 GUSTSPRING
2510 GUTHRIE
2511 GUY
2512 GUYTON
2513 GWENA
2514 GWIN
2515 GWINLY
2516 GWINNETT
2517 GYMER
2518 GYNELLE
2519 GYPNEVEE
2520 GYPSIORTHIDS
2521 GYPSUM LAND
2522 GYSTRUM

SOIL03.DAT

1 HAAR
2 HACCKE
3 HACK
4 HACKERS
5 HACKROY
6 HACKWOOD
7 HADAR
8 HADES
9 HADLEY
10 HAFLINGER
11 HAGEN
12 HAGENBARTH
13 HAGER
14 HAGERMAN
15 HAGERSTOWN
16 HAGGA
17 HAGGERTY
18 HAGSTADT
19 HAGUE
20 HAIG
21 HAIGHTS
22 HAIKU
23 HAILMAN
24 HAIRE
25 HAKKER
26 HALACAN
27 HALAQUEPTS
28 HALAWA
29 HALBERT
30 HALDER
31 HALE
32 HALEDON
33 HALEIWA
34 HALEY
35 HALF MOON
36 HALFWAY
37 HALII
38 HALIIMAILE
39 HALL
40 HALL RANCH
41 HALLANDALE
42 HALLECK
43 HALLETTSVILLE
44 HALLORAN
45 HALSEY
46 HALSO
47 HAMACER
48 HAMAKUAPOKO
49 HAMAR
50 HAMBLEN
51 HAMBONE
52 HAMBRIGHT
53 HAMBURG
54 HAMBY
55 HAMDEN
56 HAMEL
57 HAMERLY
58 HAMILTON
59 HAMLET
60 HAMLIN
61 HAMMACK
62 HAMMONTON
63 HAMPSHIRE
64 HAMPSON
65 HAMRE
66 HAMTAH
67 HANA
68 HANAKER
69 HANALEI
70 HANAMAULU
71 HANCEVILLE
72 HAND
73 HANDPAH
74 HANDRAN
75 HANDSBORO
76 HANDY
77 HANEY
78 HANFORD
79 HANGAARD
80 HANGDO
81 HANIPOE
82 HANIS
83 HANKINS
84 HANKS
85 HANKSVILLE
86 HANLON
87 HANLY
88 HANNA
89 HANNAHATCHEE
90 HANNING
91 HANOVER
92 HANS
93 HANSEL
94 HANSKA
95 HANSON
96 HANTHO
97 HANTZ
98 HAP
99 HAPGOOD
100 HAPJACK
101 HAPLAQUOLLS
102 HAPLARGIDS
103 HAPLOBOROLLS
104 HAPLOXERALFS
105 HAPLOXEROLLS
106 HAPLUDALFS
107 HAPLUMBREPTS
108 HAPLUSTOLLS
109 HAPNEY
110 HARAHAN
111 HARAHILL
112 HARANA
113 HARBORD
114 HARCANY
115 HARCO
116 HARCOT
117 HARDEMAN
118 HARDESTY
119 HARDING
120 HARDISTER
121 HARDOL
122 HARDSCRABBLE
123 HARDTRIGGER
124 HARDY
125 HARGILL
126 HARGREAVE
127 HARJO
128 HARKERS
129 HARKEY
130 HARKNESS
131 HARLAN
132 HARLEM
133 HARLESTON
134 HARLINGEN
135 HARLOW
136 HARMEHL
137 HARMONY
138 HARNEY
139 HAROL
140 HARPER
141 HARPETH
142 HARPOLE
143 HARPS
144 HARPSTER
145 HARPT
146 HARQUA
147 HARRAH
148 HARRIET
149 HARRIMAN
150 HARRINGTON
151 HARRIS
152 HARRISBURG
153 HARRISON
154 HARRISVILLE
155 HARROUN
156 HARSAN
157 HARSHA
158 HARSTINE
159 HARSTON
160 HART
161 HART CAMP
162 HARTFORD
163 HARTIG
164 HARTILL
165 HARTLAND
166 HARTLETON
167 HARTNIT
168 HARTSBURG
169 HARTSELLS
170 HARTSHORN
171 HARTVILLE
172 HARTWELL
173 HARVARD
174 HARVESTER
175 HARVEY
176 HARWOOD
177 HASKILL
178 HASKINS
179 HASSEE
180 HASSELL
181 HASTINGS
182 HAT
183 HATBORO
184 HATCH
185 HATCHET
186 HATERMUS
187 HATERTON
188 HATHAWAY
189 HATLEY
190 HATLIFF
191 HATMAKER
192 HATPEAK
193 HATTIE
194 HATTON
195 HATUR
196 HAUBSTADT
197 HAUG
198 HAUGAN
199 HAULINGS
200 HAUNCHEE
201 HAUZ
202 HAVALA
203 HAVANA
204 HAVELOCK
205 HAVEN
206 HAVERDAD
207 HAVERHILL
208 HAVERLY
209 HAVERSON
210 HAVILLAH
211 HAVINGDON
212 HAVRE
213 HAVRELON
214 HAW
215 HAWI
216 HAWICK
217 HAWKEYE
218 HAWKINS
219 HAWKSBILL
220 HAWKSPRINGS
221 HAWLEY
222 HAWSLEY
223 HAXTUN
224 HAYBOURNE
225 HAYCRIK
226 HAYDEN
227 HAYESTON
228 HAYESVILLE
229 HAYFIELD
230 HAYFORD
231 HAYHOOK
232 HAYMARKET
233 HAYMOND
234 HAYNESS
235 HAYNIE
236 HAYPRESS
237 HAYSPUR
238 HAYTER
239 HAYTI
240 HAYWIRE
241 HAYWOOD
242 HAZEL
243 HAZELAIR
244 HAZEN
245 HAZLEHURST
246 HAZLETON
247 HAZTON
248 HEADQUARTERS
249 HEAKE
250 HEALDTON
251 HEALING
252 HEARNE
253 HEATH
254 HEATHCOAT
255 HEATLY
256 HEATON
257 HEBBRONVILLE
258 HEBER
259 HEBERT
260 HEBO
261 HEBRON
262 HECATE
263 HECETA
264 HECHTMAN
265 HECKER
266 HECKISON SERIES
267 HECLA
268 HECTOR
269 HEDOES
270 HEDOX
271 HEDRICK
272 HEDVILLE
273 HEECHEE
274 HEELY
275 HEFED
276 HEFLIN
277 HEGLAR
278 HEGNE
279 HEIDEL
280 HEIDEN

281 HEIDTMAN
282 HEIGHTS
283 HEIL
284 HEIMDAL
285 HEINSAW
286 HEISETON
287 HEISLER
288 HEIST
289 HEITT
290 HEITZ
291 HEIZER
292 HELDT
293 HELEMANO
294 HELENA
295 HELENDALE
296 HELLMAN
297 HELMER
298 HELMICK
299 HELTER
300 HELVETIA
301 HELY
302 HEMBRE
303 HEMINGFORD
304 HEMPSTEAD
305 HENCO
306 HENDERSON
307 HENDRICKS
308 HENDY
309 HENEFER
310 HENHOIT
311 HENKIN
312 HENLEY
313 HENLEY, SHALLOW
314 HENLINE
315 HENMEL
316 HENNEKE
317 HENNEPIN
318 HENNESSY
319 HENNEY
320 HENNINGSEN
321 HENRIETTA
322 HENRIEVILLE
323 HENRY
324 HENSHAW
325 HENSLEY
326 HENSON
327 HEPLER
328 HEPPSIE
329 HERAKLE
330 HERBERT
331 HERBMAN
332 HERD
333 HEREFORD
334 HERKIMER
335 HERM
336 HERMANTOWN
337 HERMERING
338 HERMISTON
339 HERMITAGE
340 HERMON
341 HERMOSA
342 HERNANDEZ
343 HERNDON
344 HERO
345 HEROD
346 HERRICK
347 HERSH
348 HERSHAL
349 HERTY
350 HESCH
351 HESPER
352 HESPERIA
353 HESPERUS
354 HESSEL
355 HESSELBERG
356 HESSELTINE
357 HESSING
358 HESSLAN
359 HESSON
360 HETERWA
361 HETTINGER
362 HEUSSER
363 HEUVELTON
364 HEWITT
365 HEXT
366 HEYDER
367 HEYDLAUFF
368 HEYTOU
369 HEZEL
370 HI VISTA
371 HIARC
372 HIATT
373 HIBAR
374 HIBBING
375 HIBERNIA
376 HIBRITEN
377 HICKMAN
378 HICKORY
379 HICKS
380 HICOTA
381 HIDALGO
382 HIDATSA
383 HIDEAWAY
384 HIDEWOOD
385 HIERRO
386 HIGGINS
387 HIGGINSVILLE
388 HIGH GAP
389 HIGHAMS
390 HIGHBANK
391 HIGHCAMP
392 HIGHFIELD
393 HIGHMORE
394 HIGHPOINT
395 HIGHTOWER
396 HIGHWOOD
397 HIHIMANU
398 HIIBNER
399 HIKO PEAK
400 HIKO SPRINGS
401 HILAIRE
402 HILAND
403 HILDEBRECHT
404 HILDRETH
405 HILEA
406 HILES
407 HILGER
408 HILGRAVE
409 HILIGHT
410 HILLBRICK
411 HILLCO
412 HILLEMANN
413 HILLERY
414 HILLET
415 HILLFIELD
416 HILLGATE
417 HILLIARD
418 HILLON
419 HILLSBORO
420 HILLSDALE
421 HILLTO
422 HILLWOOD
423 HILMAR
424 HILMOE
425 HILO
426 HILOLO
427 HILT
428 HILTON
429 HINCKLEY
430 HINDES
431 HINESBURG
432 HINKER
433 HINKLE
434 HINMAN
435 HIRIDGE
436 HIRSCHDALE
437 HISEGA
438 HISER
439 HISLE
440 HISTIC CRYAQUEPT
441 HISTIC CRYAQUEPT
442 HISTIC HAPLAQUOL
443 HITILO
444 HITT
445 HIVAL
446 HIWAN
447 HIWASSEE
448 HIWOOD
449 HIXTON
450 HOADLY
451 HOBACKER
452 HOBAN
453 HOBBS
454 HOBCAW
455 HOBE
456 HOBERG
457 HOBIT
458 HOBOG
459 HOBONNY
460 HOBSON
461 HOCAR
462 HOCHHEIM
463 HOCKINSON
464 HOCKLEY
465 HODA
466 HODEDO
467 HODENPYL
468 HODGE
469 HODGINS
470 HODGSON
471 HOFFLAND
472 HOFFMANVILLE
473 HOFFSTADT
474 HOGADERO
475 HOGANSBURG
476 HOGBACK
477 HOGG
478 HOGMALAT
479 HOGRIS
480 HOH
481 HOHMANN
482 HOKO
483 HOLBROOK
484 HOLCOMB
485 HOLDAWAY
486 HOLDER
487 HOLDERMAN
488 HOLDERNESS
489 HOLDINGFORD
490 HOLDREGE
491 HOLILLIPAH
492 HOLLAND
493 HOLLANDLAKE
494 HOLLENBECK
495 HOLLINGER
496 HOLLIS
497 HOLLISTER
498 HOLLOMAN
499 HOLLOMEX
500 HOLLOWAY
501 HOLLY
502 HOLLY SPRINGS
503 HOLLYWELL
504 HOLLYWOOD
505 HOLMAN
506 HOLMDEL
507 HOLMES
508 HOLOHAN
509 HOLOMUA
510 HOLOPAW
511 HOLSINE
512 HOLSTEIN
513 HOLSTON
514 HOLT
515 HOLTER
516 HOLTLE
517 HOLTON
518 HOLTVILLE
519 HOLYOKE
520 HOMA
521 HOME CAMP
522 HOMELAKE
523 HOMELAND
524 HOMER
525 HOMESTAKE
526 HOMESTEAD
527 HOMEWOOD
528 HOMME
529 HOMOSASSA
530 HONAUNAU
531 HONCUT
532 HONDALE
533 HONDOHO
534 HONEOYE
535 HONEYGROVE
536 HONEYVILLE
537 HONKER
538 HONN
539 HONOBIA
540 HONOKAA
541 HONOLUA
542 HONOMANU
543 HONONEGAH
544 HONOULIULI
545 HONTAS
546 HONTOON
547 HONUAULU
548 HOOD
549 HOODLE
550 HOODOO
551 HOODSPORT
552 HOOGDAL
553 HOOKS
554 HOOLEHUA
555 HOOLY
556 HOOPAL
557 HOOPER
558 HOOPESTON
559 HOOPLITE
560 HOOSAN
561 HOOSEGOW
562 HOOSIC
563 HOOSIERVILLE
564 HOOT
565 HOOTEN
566 HOPCO
567 HOPDRAW
568 HOPEKA
569 HOPKINS
570 HOPLAND
571 HOPLEY
572 HOPSONVILLE
573 HOQUIAM
574 HORD
575 HOREB
576 HORNELL
577 HORNING
578 HORNITOS
579 HORNSBY
580 HORNSVILLE

581 HORROCKS
582 HORSECAMP
583 HORSERIDGE
584 HORSESHOE
585 HORSETHIEF
586 HORSLEY
587 HORTONVILLE
588 HOSKIN
589 HOSKINNINI
590 HOSLEY
591 HOSMER
592 HOSPAL
593 HOSSICK
594 HOSTAGE
595 HOT LAKE
596 HOTAW
597 HOTCREEK
598 HOTEL
599 HOTSPRINGS
600 HOUDEK
601 HOUGH
602 HOUGHTON
603 HOUGHTONVILLE
604 HOUK
605 HOULA
606 HOULKA
607 HOURGLASS
608 HOUSE MOUNTAIN
609 HOUSER
610 HOUSTAKE
611 HOUSTON
612 HOUSTON BLACK
613 HOVDE
614 HOVEN
615 HOVENWEEP
616 HOVERT
617 HOVEY
618 HOWARD
619 HOWCAN
620 HOWCREE
621 HOWE
622 HOWELL
623 HOWLAND
624 HOWSON
625 HOYE
626 HOYLETON
627 HOYPUS
628 HOYTVILLE
629 HUALAPAI
630 HUB
631 HUBBARD
632 HUBBARDTON
633 HUBBELL
634 HUBERLY
635 HUBERT
636 HUBLERSBURG
637 HUCKLEBERRY
638 HUDSON
639 HUECO
640 HUEL
641 HUENEME
642 HUERFANO
643 HUEY
644 HUFFINE
645 HUFFMAN
646 HUFFTON
647 HUGGINS
648 HUGHES
649 HUGHESVILLE
650 HUGO
651 HUGUSTON
652 HUICHICA
653 HUIKAU
654 HUKILL
655 HULETT
656 HULLS
657 HULLT
658 HULUA
659 HUM
660 HUMACAO
661 HUMATAS
662 HUMBARGER
663 HUMBIG
664 HUMBIRD
665 HUMBOLDT
666 HUMDUN
667 HUMESTON
668 HUMIC CRYAQUEPTS
669 HUMKER
670 HUMMINGTON
671 HUMPHREYS
672 HUMPTULIPS
673 HUMSKEL
674 HUN
675 HUNCHBACK
676 HUNEWILL
677 HUNGRY
678 HUNNTON
679 HUNTERS
680 HUNTERSVILLE
681 HUNTIMER
682 HUNTING
683 HUNTINGTON
684 HUNTMOUNT
685 HUNTSBURG
686 HUNTSVILLE
687 HUPP
688 HURDS
689 HURLBUT
690 HURLEY
691 HURRICANE
692 HURST
693 HUSE
694 HUSKA
695 HUSSA
696 HUSSMAN
697 HUSUM
698 HUTCHINSON
699 HUTSON
700 HUTT
701 HUTTON
702 HUXLEY
703 HUYSINK
704 HYANNIS
705 HYAS
706 HYATTVILLE
707 HYDE
708 HYDER
709 HYDRANDEPTS
710 HYDRAQUENTS
711 HYDRO
712 HYE
713 HYLOC
714 HYMAS
715 HYRUM
716 HYSHAM
717 HYZEN
718 IAO
719 IBERIA
720 ICARIA
721 ICENE
722 ICESLEW
723 ICHBOD
724 ICHETUCKNEE
725 ICICLE
726 IDA
727 IDABEL
728 IDAMONT
729 IDEE
730 IDLEWILD
731 IDMON
732 IGDELL
733 IGERT
734 IGNACIO
735 IGO
736 IGUALDAD
737 IHLEN
738 IJAM
739 ILACHETOMEL
740 ILDEFONSO
741 ILES
742 ILIFF
743 ILIILI
744 ILION
745 ILLABOT
746 ILLER
747 ILTON
748 ILWACO
749 IMA
750 IMBLER
751 IMLAY
752 IMMIG
753 IMMOKALEE
754 IMOGENE
755 IMONIL
756 IMPACT
757 IMPERIAL
758 INAVALE
759 INCELL
760 INCHAU
761 INCY
762 INDART
763 INDEX
764 INDIAHOMA
765 INDIAN CREEK
766 INDIANO
767 INDIANOLA
768 INDIO
769 INDUS
770 INEZ
771 INGALLS
772 INGENIO
773 INGERSOLL
774 INGRAM
775 INKLER
776 INKOM
777 INKOSR
778 INKS
779 INKSTER
780 INLOW
781 INMACHUK
782 INMAN
783 INMO
784 INNINGER
785 INPENDENCE
786 INSAK
787 INSKIP
788 INSULA
789 INTERIOR
790 INTON
791 INVERNESS
792 INVERSHIEL
793 INVILLE
794 IO
795 IOLEAU
796 IONA
797 IONIA
798 IOSCO
799 IOSEPA
800 IOTLA
801 IPAGE
802 IPANO
803 IPAVA
804 IPISH
805 IPSON
806 IPSUT
807 IPSWICH
808 IRA
809 IRAAN
810 IREDELL
811 IRELAND
812 IRETEBA
813 IRIGUL
814 IRIM
815 IRMULCO
816 IROCK
817 IRON BLOSSOM
818 IRON MOUNTAIN
819 IRON RIVER
820 IRONCO
821 IRONDALE
822 IRONTON
823 IROQUOIS
824 IRRAWADDY
825 IRRIGON
826 IRVINE
827 IRVINGTON
828 IRWIN
829 ISAAC
830 ISABELLA
831 ISAN
832 ISANTI
833 ISBELL
834 ISHI PISHI
835 ISHPEMING
836 ISKNAT
837 ISLAND
838 ISLES
839 ISLOTE
840 ISOLDE
841 ISOM
842 ISTER
843 ISTOKPOGA
844 ITANO
845 ITASCA
846 ITAT
847 ITCA
848 ITHACA
849 ITSWOOT
850 IUKA
851 IVA
852 IVAN
853 IVANELL
854 IVER
855 IVES
856 IVINS
857 IYRES
858 IZAGORA
859 IZAR
860 IZEE
861 IZO
862 IZUSER
863 JABU
864 JACAGUAS
865 JACANA
866 JACEE
867 JACINTO
868 JACK CREEK
869 JACKET
870 JACKLAND
871 JACKMAN
872 JACKNIFE
873 JACKPORT
874 JACKS
875 JACKSON
876 JACKTONE
877 JACOB
878 JACOBSEN
879 JACOBY
880 JACOT

881 JACQUES
882 JACQUITH
883 JACRATZ
884 JACWIN
885 JADIS
886 JAFA
887 JAGUEYES
888 JAL
889 JALMAR
890 JAMES
891 JAMES CANYON
892 JAMESTON
893 JANISE
894 JANSEN
895 JANUDE
896 JARAB
897 JARBOE
898 JARDIN
899 JAREALES
900 JARITA
901 JARMILLO
902 JAROLA
903 JAROSO
904 JARRE
905 JARRON
906 JARVIS
907 JASCO
908 JASON
909 JASPER
910 JAUCAS
911 JAVA
912 JAWBONE
913 JAY
914 JAYAR
915 JAYBEE
916 JAYEL
917 JAYEM
918 JAYNES
919 JEAGER
920 JEAN
921 JEAN LAKE
922 JEANERETTE
923 JEBO
924 JEDD
925 JEDDITO
926 JEDDO
927 JEFFERS
928 JEFFERSON
929 JEFFREY
930 JEKLEY
931 JEMEZ
932 JENA
933 JENKINS
934 JENKINSON
935 JENNESS
936 JENNINGS
937 JENNY
938 JENOR
939 JERAG
940 JERAULD
941 JERICHO
942 JEROME
943 JERRY
944 JERRY VARIANT
945 JERRYSLU
946 JERU
947 JERVAL
948 JESSE CAMP
949 JESSO
950 JESSUP
951 JETSTER
952 JETT
953 JEVETS
954 JEWETT
955 JIGGS
956 JILSON
957 JIM
958 JIMBO
959 JIMEK
960 JIMENEZ
961 JIMLAKE
962 JIMSAGE
963 JIMTOWN
964 JIPPER
965 JIVAS
966 JOACHEM
967 JOB
968 JOBOS
969 JOBPEAK
970 JOCITY
971 JOCKO
972 JODERO
973 JOEL
974 JOENEY
975 JOES
976 JOEVAR
977 JOHNNIE
978 JOHNS
979 JOHNSBURG
980 JOHNSON
981 JOHNSTON
982 JOHNSWOOD
983 JOICE
984 JOINER
985 JOKODOWSKI
986 JOLAN
987 JOLIET
988 JONALE
989 JONAS
990 JONATHAN
991 JONCA
992 JONDA
993 JONES
994 JONESVILLE
995 JOPLIN
996 JOPPA
997 JORDAN
998 JORGE
999 JORY
1000 JOSBURG
1001 JOSEPHINE
1002 JOSHUA
1003 JOSIE
1004 JOSLIN
1005 JOURDANTON
1006 JOWEC
1007 JOY
1008 JUAB
1009 JUANA DIAZ
1010 JUBILEE
1011 JUDA
1012 JUDD
1013 JUDELL
1014 JUDICE
1015 JUDITH
1016 JUDKINS
1017 JUDSON
1018 JUDY
1019 JUG
1020 JUGET
1021 JUGHANDLE
1022 JUGSON
1023 JULES
1024 JULESBURG
1025 JULIN
1026 JUMBO
1027 JUMPE
1028 JUMPER
1029 JUMPMORE
1030 JUMPOFF
1031 JUNCAL
1032 JUNCOS
1033 JUNCTION
1034 JUNEAU
1035 JUNG
1036 JUNIPERBUTE
1037 JUNIPERO
1038 JUNIUS
1039 JUNKETT
1040 JUNO
1041 JUNQUITOS
1042 JUPITER
1043 JURA
1044 JURVANNAH
1045 JUSTESEN
1046 JUSTIN
1047 JUVA
1048 JUVAN
1049 KAALUALU
1050 KACHEMAK
1051 KACHESS
1052 KADE
1053 KADLETZ
1054 KADOKA
1055 KAENA
1056 KAFING
1057 KAHALUU
1058 KAHANA
1059 KAHANUI
1060 KAHLER
1061 KAHLOTUS
1062 KAHOLA
1063 KAHUA
1064 KAIDERS
1065 KAILUA
1066 KAIMU
1067 KAINALIU
1068 KAIPOIOI
1069 KAIWIKI
1070 KALAE
1071 KALALOCH
1072 KALAMA
1073 KALAMAZOO
1074 KALAPA
1075 KALAUPAPA
1076 KALEETAN
1077 KALIFONSKY
1078 KALIGA
1079 KALIHI
1080 KALISPELL
1081 KALKASKA
1082 KALLIO
1083 KALMARVILLE
1084 KALMIA
1085 KALOKO
1086 KALONA
1087 KALSIN
1088 KALSTED
1089 KAMACK
1090 KAMAKOA
1091 KAMAN
1092 KAMAOA
1093 KAMAOLE
1094 KAMAY
1095 KAMELA
1096 KAMIE
1097 KAMPVILLE
1098 KAMRAR
1099 KANAKA
1100 KANAPAHA
1101 KANARANZI
1102 KANARRA
1103 KANASKAT
1104 KANAWHA
1105 KANDALY
1106 KANDIK
1107 KANDOTA
1108 KANE
1109 KANEBREAK
1110 KANEOHE
1111 KANEPUU
1112 KANER
1113 KANG
1114 KANGAS
1115 KANID
1116 KANIKSU
1117 KANIMA
1118 KANKAKEE
1119 KANLEE
1120 KANONA
1121 KANOSH
1122 KANTISHNA
1123 KANUTCHAN
1124 KANZA
1125 KAPAA
1126 KAPAPALA
1127 KAPIN
1128 KAPOD
1129 KAPOWSIN
1130 KAPTURE
1131 KAPUHIKANI
1132 KARAMIN
1133 KARANKAWA
1134 KARCAL
1135 KARDE
1136 KARLAN
1137 KARLIN
1138 KARLO
1139 KARLSBURG
1140 KARLSTAD
1141 KARLUK
1142 KARMA
1143 KARNAK
1144 KARNES
1145 KARPP
1146 KARRO
1147 KARS
1148 KARSHNER
1149 KARTAR
1150 KASEBERG
1151 KASHWITNA
1152 KASILOF
1153 KASKI
1154 KASOTA
1155 KASSLER
1156 KASSON
1157 KATAMA
1158 KATEMCY
1159 KATHER
1160 KATO
1161 KATSEANES
1162 KATULA
1163 KATY
1164 KAUDER
1165 KAUFMAN
1166 KAUKAUNA
1167 KAUPO
1168 KAUPPI
1169 KAVETT
1170 KAWAIHAE
1171 KAWAIHAPAI
1172 KAWBAWGAM
1173 KAWICH
1174 KAWKAWLIN
1175 KAYO
1176 KEAAU
1177 KEAHUA
1178 KEALAKEKUA
1179 KEALIA
1180 KEARL

1181 KEARNS
1182 KEARSARGE
1183 KEATING
1184 KEAUKAHA
1185 KEAWAKAPU
1186 KEBLER
1187 KECH
1188 KECKO
1189 KECKSROAD
1190 KEDA
1191 KEDDIE
1192 KEDRON
1193 KEE
1194 KEECHELUS
1195 KEECHI
1196 KEEFERS
1197 KEEI
1198 KEEKEE
1199 KEEL
1200 KEELDAR
1201 KEELE
1202 KEELINE
1203 KEENE
1204 KEENO
1205 KEESE
1206 KEESEHA
1207 KEETER
1208 KEEWATIN
1209 KEG
1210 KEGEL
1211 KEGONSA
1212 KEHENA
1213 KEHOE
1214 KEIGLEY
1215 KEISER
1216 KEITH
1217 KEITHVILLE
1218 KEKAHA
1219 KEKAKE
1220 KEKAWAKA
1221 KELK
1222 KELLER
1223 KELLY
1224 KELSEY
1225 KELSO
1226 KELTNER
1227 KELTYS
1228 KELVIN
1229 KEMAH
1230 KEMAN
1231 KEMMERER
1232 KEMOO
1233 KEMP
1234 KEMPSVILLE
1235 KENAI
1236 KENANSVILLE
1237 KENDAIA
1238 KENDALL
1239 KENDALLVILLE
1240 KENDRICK
1241 KENESAW
1242 KENMOOR
1243 KENN
1244 KENNAN
1245 KENNEBEC
1246 KENNEDY
1247 KENNER
1248 KENNEWICK
1249 KENNEY
1250 KENNEY LAKE
1251 KENO
1252 KENOMA
1253 KENSAL
1254 KENSETT
1255 KENSPUR
1256 KENT
1257 KENUSKY
1258 KENYON
1259 KEO
1260 KEOKUK
1261 KEOMAH
1262 KEOTA
1263 KEOWNS
1264 KEPLER
1265 KERBER
1266 KERBY
1267 KERL
1268 KERMIT
1269 KERNAN
1270 KERRDAM
1271 KERRICK
1272 KERRVILLE
1273 KERSHAW
1274 KERSICK
1275 KERSTON
1276 KERT
1277 KESSLER
1278 KESSON
1279 KESTERSON
1280 KESWICK
1281 KETCHLY
1282 KETCHUM
1283 KETONA
1284 KETTENBACH
1285 KETTLE
1286 KETTLEMAN
1287 KETTNER
1288 KEUTERVILLE
1289 KEVANTON
1290 KEVIN
1291 KEWAUNEE
1292 KEWEENAW
1293 KEYA
1294 KEYES
1295 KEYNER
1296 KEYPORT
1297 KEYSTONE
1298 KEZAN
1299 KEZAR
1300 KIAKUS
1301 KIAWAH
1302 KIBBIE
1303 KIBESILLAH
1304 KICKAPOO
1305 KICKERVILLE
1306 KIDD
1307 KIDDER
1308 KIDMAN
1309 KIEHL
1310 KIESEL
1311 KIETZKE
1312 KIEV
1313 KIKONI
1314 KILAGA
1315 KILARC
1316 KILAUEA
1317 KILBURN
1318 KILCHIS
1319 KILDOR
1320 KILFOIL
1321 KILGORE
1322 KILKENNY
1323 KILLBUCK
1324 KILLDUFF
1325 KILLEY
1326 KILLPACK
1327 KILMANAGH
1328 KILMER
1329 KILMERQUE
1330 KILN
1331 KILOA
1332 KILOHANA
1333 KILOWAN
1334 KILWINNING
1335 KIM
1336 KIMAMA
1337 KIMBALL
1338 KIMBERLINA
1339 KIMBERLY
1340 KIMBROUGH
1341 KIMMERLING
1342 KIMO
1343 KINA
1344 KINCHELOE
1345 KINCO
1346 KINDER
1347 KINDIG
1348 KINDY
1349 KINESAVA
1350 KINGFISHER
1351 KINGHORN
1352 KINGILE
1353 KINGINGHAM
1354 KINGMAN
1355 KINGMONT
1356 KINGS
1357 KINGSBURY
1358 KINGSDOWN
1359 KINGSLAND
1360 KINGSLEY
1361 KINGSTON
1362 KINGSVILLE
1363 KINGTAIN
1364 KINKEAD
1365 KINKEL
1366 KINKORA
1367 KINMAN
1368 KINNEAR
1369 KINNEY
1370 KINROSS
1371 KINSTON
1372 KINTA
1373 KINTON
1374 KINZEL
1375 KIOMATIA
1376 KIONA
1377 KIOTE
1378 KIPER
1379 KIPLING
1380 KIPPEN
1381 KIPSON
1382 KIRBY
1383 KIRBYVILLE
1384 KIRK
1385 KIRKENDALL
1386 KIRKHAM
1387 KIRKLAND
1388 KIRKSEY
1389 KIRKVILLE
1390 KIRLEY
1391 KIRTLEY
1392 KIRVIN
1393 KISATCHIE
1394 KISHONA
1395 KISRING
1396 KISSICK
1397 KISTIRN
1398 KITCHELL
1399 KITCHEN CREEK
1400 KITI
1401 KITSAP
1402 KITTERLL
1403 KITTITAS
1404 KITTREDGE
1405 KITTSON
1406 KIVA
1407 KIWANIS
1408 KIZHUYAK
1409 KJAR
1410 KLABER
1411 KLADNICK
1412 KLAMATH
1413 KLAPATCHE
1414 KLAUS
1415 KLAWASI
1416 KLAWATTI
1417 KLAWHOP
1418 KLAYENT
1419 KLECKNER
1420 KLEINBUSH
1421 KLEJ
1422 KLICKER
1423 KLICKITAT
1424 KLICKSON
1425 KLINE
1426 KLINESVILLE
1427 KLINGER
1428 KLONE
1429 KLOOCHMAN
1430 KLOOTCH
1431 KLOOTCHIE
1432 KLOTEN
1433 KLUG
1434 KLUM
1435 KLUMP
1436 KLUTINA
1437 KNAPKE
1438 KNAPPA
1439 KNAPPTON
1440 KNEELAND
1441 KNEP
1442 KNICKERBOCKER
1443 KNIFFIN
1444 KNIGHT
1445 KNIK
1446 KNIKLIK
1447 KNIPPA
1448 KNOB HILL
1449 KNOBTOP
1450 KNOCO
1451 KNOKE
1452 KNOLLE
1453 KNOSS
1454 KNOTT
1455 KNOWLES
1456 KNOX
1457 KNULL
1458 KNUTSEN
1459 KOBAR
1460 KOBEH
1461 KOBEL
1462 KOCH
1463 KODAK
1464 KODIAK
1465 KODRA
1466 KOEHLER
1467 KOELE
1468 KOEPKE
1469 KOERLING
1470 KOERTH
1471 KOETHER
1472 KOFA
1473 KOGISH
1474 KOHALA
1475 KOKAN
1476 KOKEE
1477 KOKERNOT
1478 KOKO
1479 KOKOKAHI
1480 KOKOMO

1481 KOLAR
1482 KOLBERG
1483 KOLEKOLE
1484 KOLIN
1485 KOLLS
1486 KOLLUTUK
1487 KOLOA
1488 KOLOB
1489 KOLOKOLO
1490 KOLOMOKI
1491 KOMO
1492 KONA
1493 KONAWA
1494 KONERT
1495 KONNER
1496 KONOCTI
1497 KONSIL
1498 KOOLAU
1499 KOONICH
1500 KOONTZ
1501 KOOSHAREM
1502 KOOSKIA
1503 KOOTENAI
1504 KOPIE
1505 KOPPERL
1506 KOPPES
1507 KORCHEA
1508 KORENT
1509 KORNMAN
1510 KOROBAGO
1511 KORONIS
1512 KORTTY
1513 KOSCIUSKO
1514 KOSETH
1515 KOSSE
1516 KOSSUTH
1517 KOSZTA
1518 KOTO
1519 KOTZMAN
1520 KOURY
1521 KOVICH
1522 KOYEN
1523 KOYNIK
1524 KOYUKUK
1525 KRADE
1526 KRAKON
1527 KRAM
1528 KRANSKI
1529 KRANZBURG
1530 KRATKA
1531 KRAUSE
1532 KREAMER
1533 KREBS
1534 KREM
1535 KREMLIN
1536 KRESSON
1537 KREYENHAGEN
1538 KRIER
1539 KRIEST
1540 KRONEN
1541 KRUEGER
1542 KRUM
1543 KRUSE
1544 KUBE
1545 KUBLER
1546 KUBLI
1547 KUCERA
1548 KUCK
1549 KUDLAC
1550 KUHL
1551 KUKAIAU
1552 KULA
1553 KULLIT
1554 KULSHAN
1555 KUMA
1556 KUNATON
1557 KUNAYOSH
1558 KUNIA
1559 KUNUWEIA
1560 KUNZ
1561 KUPREANOF
1562 KUREB
1563 KURO
1564 KURTH
1565 KURTZ
1566 KUSKOKWIM
1567 KUSLINA
1568 KUTCH
1569 KUTLER
1570 KUY
1571 KVICHAK
1572 KWEO
1573 KYBURZ
1574 KYDAKA
1575 KYLE
1576 KYLER
1577 LA BRIER
1578 LA FARGE
1579 LA FONDA
1580 LA GRANDE
1581 LA HOGUE
1582 LA LANDE
1583 LA PALMA
1584 LA POSTA
1585 LA PRAIRIE
1586 LA ROSE
1587 LABENZO
1588 LABETTE
1589 LABISH
1590 LABISKI
1591 LABOU
1592 LABOUNTY
1593 LABRE
1594 LABSHAFT
1595 LABU
1596 LABUCK
1597 LACAMAS
1598 LACERDA
1599 LACHAPELLA
1600 LACITA
1601 LACKAWANNA
1602 LACLEDE
1603 LACONNER
1604 LACOOCHEE
1605 LACOSTE
1606 LACOTA
1607 LACRESCENT
1608 LACY
1609 LADD
1610 LADELLE
1611 LADERLY
1612 LADNER
1613 LADOGA
1614 LADUE
1615 LADYSMITH
1616 LAFE
1617 LAFITTE
1618 LAG
1619 LAGLORIA
1620 LAGNAF
1621 LAGONDA
1622 LAGRANGE
1623 LAGROSS
1624 LAGUNITA
1625 LAHAINA
1626 LAHONTAN
1627 LAHRITY
1628 LAIDIG
1629 LAIDLAW
1630 LAIL
1631 LAIRD
1632 LAIRDSVILLE
1633 LAJARA
1634 LAJITAS
1635 LAKE
1636 LAKE CHARLES
1637 LAKE CREEK
1638 LAKE JANEE
1639 LAKEHELEN
1640 LAKEHURST
1641 LAKELAND
1642 LAKEMONT
1643 LAKEPORT
1644 LAKESHORE
1645 LAKESIDE
1646 LAKESOL
1647 LAKETON
1648 LAKEVIEW
1649 LAKEWIN
1650 LAKEWOOD
1651 LAKI
1652 LAKIN
1653 LAKOA
1654 LAKOMA
1655 LAKRIDGE
1656 LALAAU
1657 LALINDA
1658 LALLIE
1659 LAM
1660 LAMA
1661 LAMANGA
1662 LAMAR
1663 LAMARSH
1664 LAMARTINE
1665 LAMATH
1666 LAMAWA
1667 LAMBERT
1668 LAMBETH
1669 LAMBMAN
1670 LAMBRING
1671 LAMEDEER
1672 LAMINGTON
1673 LAMKIN
1674 LAMO
1675 LAMOILLE
1676 LAMONDI
1677 LAMONI
1678 LAMONT
1679 LAMONTA
1680 LAMOOSE
1681 LAMOTTE
1682 LAMOURE
1683 LAMPHIER
1684 LAMPSHIRE
1685 LAMSON
1686 LANADA
1687 LANARK
1688 LANCASTER
1689 LANCE
1690 LAND
1691 LANDCO
1692 LANDER
1693 LANDES
1694 LANDLOW
1695 LANDMAN
1696 LANE
1697 LANEY
1698 LANG
1699 LANGFORD
1700 LANGHEI
1701 LANGLADE
1702 LANGLOIS
1703 LANGOLA
1704 LANGRELL
1705 LANGSPRING
1706 LANGSTON
1707 LANGTRY
1708 LANIER
1709 LANIGER
1710 LANKBUSH
1711 LANKIN
1712 LANKTREE
1713 LANOAK
1714 LANSDALE
1715 LANSDOWNE
1716 LANSING
1717 LANTERN
1718 LANTIS
1719 LANTON
1720 LANTONIA
1721 LANTRY
1722 LANTZ
1723 LANVER
1724 LANYON
1725 LAP
1726 LAPARITA
1727 LAPDUN
1728 LAPED
1729 LAPEER
1730 LAPHAM
1731 LAPINE
1732 LAPLATTA
1733 LAPON
1734 LAPORTE
1735 LAPOSA
1736 LARAND
1737 LARCHMOUNT
1738 LARDELL
1739 LAREDO
1740 LARES
1741 LARGO
1742 LARIAT
1743 LARIM
1744 LARIMER
1745 LARIOSCAMP
1746 LARKIN
1747 LARKSON
1748 LARMINE
1749 LAROQUE
1750 LAROSE
1751 LARRUPIN
1752 LARRY
1753 LARSON
1754 LARTON
1755 LARUE
1756 LARUSH
1757 LARVIE
1758 LAS
1759 LAS ANIMAS
1760 LAS FLORES
1761 LAS LUCAS
1762 LAS POSAS
1763 LAS VEGAS
1764 LASA
1765 LASALLE
1766 LASAUSES
1767 LASIL
1768 LASKA
1769 LASSEL
1770 LASSEN
1771 LASSITER
1772 LASTANCE
1773 LATAH
1774 LATAHCO
1775 LATANIER
1776 LATCH
1777 LATENE
1778 LATES
1779 LATHAM
1780 LATHER

1781 LATHROP
1782 LATIGO
1783 LATINA
1784 LATIUM
1785 LATOM
1786 LATONIA
1787 LATOUCHE
1788 LATOURELL
1789 LATTAS
1790 LATTY
1791 LAUDERDALE
1792 LAUDERHILL
1793 LAUFER
1794 LAUGENOUR
1795 LAUGHLIN
1796 LAUMAIA
1797 LAUREL
1798 LAURELWOOD
1799 LAUREN
1800 LAURENTZEN
1801 LAVA FLOWS
1802 LAVACREEK
1803 LAVALLEE
1804 LAVATE
1805 LAVEAGA
1806 LAVEEN
1807 LAVELDO
1808 LAVERKIN
1809 LAVIC
1810 LAVINA
1811 LAWAI
1812 LAWET
1813 LAWLER
1814 LAWNDALE
1815 LAWNWOOD
1816 LAWRENCE
1817 LAWRENCEVILLE
1818 LAWSHE
1819 LAWSON
1820 LAWTHER
1821 LAWTON
1822 LAWYER
1823 LAX
1824 LAXAL
1825 LAXTON
1826 LAYCOCK
1827 LAYTON
1828 LAYVIEW
1829 LAZEAR
1830 LAZMAN
1831 LE BAR
1832 LE SUEUR
1833 LEA
1834 LEADER
1835 LEADPOINT
1836 LEADVALE
1837 LEADVILLE
1838 LEAF
1839 LEAFU
1840 LEAGUEVILLE
1841 LEAKSVILLE
1842 LEAL
1843 LEALANDIC
1844 LEANNA
1845 LEANTO
1846 LEAPS
1847 LEATHAM
1848 LEATHERMAN
1849 LEAVENWORTH
1850 LEAVITT
1851 LEAVITTVILLE
1852 LEBAM
1853 LEBANON
1854 LEBEAU
1855 LEBEC
1856 LEBO
1857 LEBSACK
1858 LECK KILL
1859 LEDFORD
1860 LEDGEFORK
1861 LEDMOUNT
1862 LEDOW
1863 LEDRU
1864 LEDUB
1865 LEDUCK
1866 LEDWITH
1867 LEE
1868 LEEBENCH
1869 LEEDS
1870 LEEFIELD
1871 LEEKO
1872 LEELANAU
1873 LEEMONT
1874 LEEPER
1875 LEERAY
1876 LEESBURG
1877 LEETONIA
1878 LEEVAN
1879 LEFOR
1880 LEGAULT
1881 LEGGETT
1882 LEGLER
1883 LEGORE
1884 LEHEW
1885 LEHIGH
1886 LEHMANS
1887 LEHR
1888 LEICESTER
1889 LEIDL
1890 LEIGHCAN
1891 LEILEHUA
1892 LEISY
1893 LELA
1894 LELAND
1895 LEMAH
1896 LEMCO
1897 LEMERT
1898 LEMETA
1899 LEMING
1900 LEMM
1901 LEMOLO
1902 LEMOND
1903 LEMONEX
1904 LEMOORE
1905 LEN
1906 LENA
1907 LENAPAH
1908 LENAWEE
1909 LENBERG
1910 LENNEP
1911 LENOIR
1912 LENZ
1913 LENZBURG
1914 LEO
1915 LEOLA
1916 LEON
1917 LEONARD
1918 LEONARDO
1919 LEONARDTOWN
1920 LEONI
1921 LERDAL
1922 LERDO
1923 LEROY
1924 LERROW
1925 LESHARA
1926 LESHO
1927 LESLIE
1928 LESON
1929 LESPATE
1930 LESTER
1931 LESWILL
1932 LETA
1933 LETCHER
1934 LETHA
1935 LETHENT
1936 LETNEY
1937 LETON
1938 LETORT
1939 LETRI
1940 LETTIA
1941 LEVASY
1942 LEVEE
1943 LEVELTON
1944 LEVERETT
1945 LEVIATHAN
1946 LEVIS
1947 LEVY
1948 LEW
1949 LEWIS
1950 LEWISBERRY
1951 LEWISBURG
1952 LEWISTON
1953 LEWISVILLE
1954 LEWKALB
1955 LEX
1956 LEXINGTON
1957 LEXTON
1958 LEYBA
1959 LEYDEN
1960 LIBBINGS
1961 LIBEG
1962 LIBERAL
1963 LIBORY
1964 LIBRARY
1965 LIBUSE
1966 LICHA
1967 LICK
1968 LICKDALE
1969 LICKING
1970 LICKSKILLET
1971 LIDDELL
1972 LIDDIEVILLE
1973 LIDY
1974 LIEN
1975 LIGGET
1976 LIGHTNING
1977 LIGNUM
1978 LIGON
1979 LIGURTA
1980 LIHEN
1981 LIHUE
1982 LIKES
1983 LILAH
1984 LILBERT
1985 LILBOURN
1986 LILLINGS
1987 LILLINGTON
1988 LILY
1989 LIMA
1990 LIMBER
1991 LIMEKILN
1992 LIMERICK
1993 LIMERIDGE
1994 LIMON
1995 LIMONES
1996 LIMPIA
1997 LINCO
1998 LINCOLN
1999 LINDAAS
2000 LINDALE
2001 LINDELL
2002 LINDEN
2003 LINDER
2004 LINDLEY
2005 LINDRITH
2006 LINDSIDE
2007 LINDSTROM
2008 LINDY
2009 LINE
2010 LINEVILLE
2011 LINGANORE
2012 LINHART
2013 LININGER
2014 LINK
2015 LINKER
2016 LINKVILLE
2017 LINNE
2018 LINNET
2019 LINNEUS
2020 LINO
2021 LINOYER
2022 LINROSE
2023 LINSLAW
2024 LINT
2025 LINTON
2026 LINVELDT
2027 LINVILLE
2028 LINWELL
2029 LINWOOD
2030 LIPAN
2031 LIPKE
2032 LIPPINCOTT
2033 LIPPITT
2034 LIRIOS
2035 LISADE
2036 LISAM
2037 LISBON
2038 LISCO
2039 LISCOMB
2040 LISK
2041 LISMAS
2042 LISMORE
2043 LITCHFIELD
2044 LITHGOW
2045 LITHGOW, DEEP
2046 LITHIC CRYOBOROL
2047 LITHIC HAPLOXERO
2048 LITHIC TORRIORTH
2049 LITHIC UDORTHENT
2050 LITHIC UMBRIC VI
2051 LITHIC USTIC TOR
2052 LITHIC USTIC TOR
2053 LITHIC USTORTHEN
2054 LITHIC VITRANDEP
2055 LITHIC XEROCHREP
2056 LITHIC XEROCRHEP
2057 LITHIC XERORTHEN
2058 LITIMBER
2059 LITLE
2060 LITRO
2061 LITTLE HORN
2062 LITTLE POLE
2063 LITTLE WOOD
2064 LITTLEBEAR
2065 LITTLEJOHN
2066 LITTLETON
2067 LITTSAN
2068 LITZ
2069 LIV
2070 LIVENGOOD
2071 LIVEOAK
2072 LIVERMORE
2073 LIVIA
2074 LIVINGSTON
2075 LIVONA
2076 LIZE
2077 LIZZANT
2078 LLANOS
2079 LOARC
2080 LOBDELL

2081 LOBELVILLE
2082 LOBERG
2083 LOBERT
2084 LOBITOS
2085 LOBO
2086 LOBURN
2087 LOCANE
2088 LOCEY
2089 LOCHLOOSA
2090 LOCHSA
2091 LOCKE
2092 LOCKERBY
2093 LOCKHART
2094 LOCKMAN
2095 LOCKPORT
2096 LOCKTON
2097 LOCKWOOD
2098 LOCO
2099 LOCODA
2100 LOCODA SERIES
2101 LOCUST
2102 LODALLEY
2103 LODAR
2104 LODE
2105 LODI
2106 LODO
2107 LOFFTUS
2108 LOFTON
2109 LOGAN
2110 LOGDELL
2111 LOGGERT
2112 LOGHOUSE
2113 LOGY
2114 LOHLER
2115 LOHMILLER
2116 LOHNES
2117 LOHSMAN
2118 LOIRE
2119 LOKEN
2120 LOKERN
2121 LOKOSEE
2122 LOLAK
2123 LOLALITA
2124 LOLEKAA
2125 LOLO
2126 LOLON
2127 LOLOPEAK
2128 LOMA
2129 LOMAKI
2130 LOMALTA
2131 LOMART
2132 LOMAX
2133 LOMETA
2134 LOMIRA
2135 LOMITAS
2136 LOMOINE
2137 LOMOND
2138 LOMPICO
2139 LONCAN
2140 LONDO
2141 LONDONDERRY
2142 LONE
2143 LONE ROCK
2144 LONEPINE
2145 LONERIDGE
2146 LONESTAR
2147 LONETREE
2148 LONEWOOD
2149 LONGCREEK
2150 LONGDOE
2151 LONGFORD
2152 LONGJIM
2153 LONGLOIS
2154 LONGMARE
2155 LONGMONT
2156 LONGRIE
2157 LONGVAL
2158 LONGVIEW
2159 LONIGAN
2160 LONJON
2161 LONNA
2162 LONOKE
2163 LONTI
2164 LOOKINGGLASS
2165 LOOKOUT
2166 LOOMER
2167 LOOMIS
2168 LOPER
2169 LOPEZ
2170 LOPWASH
2171 LORACK
2172 LORADALE
2173 LORAIN
2174 LORAN
2175 LORDSTOWN
2176 LOREAUVILLE
2177 LORELLA
2178 LORENA
2179 LORENZO
2180 LORETTO
2181 LORING
2182 LORMAN
2183 LORTA
2184 LOS ALAMOS
2185 LOS BANOS
2186 LOS GATOS
2187 LOS GUINEOS
2188 LOS OSOS
2189 LOS ROBLES
2190 LOS TANOS
2191 LOSANTVILLE
2192 LOSEE
2193 LOSTINE
2194 LOSTPOINT
2195 LOSTSPRING
2196 LOSTVALLEY
2197 LOSTWELLS
2198 LOTHAIR
2199 LOTT
2200 LOUDERBACK
2201 LOUDON
2202 LOUDONVILLE
2203 LOUELLA
2204 LOUGHBORO
2205 LOUGHTY
2206 LOUIE
2207 LOUIN
2208 LOUISA
2209 LOUISBURG
2210 LOUP
2211 LOUPLOUP
2212 LOURDES
2213 LOUSCOT
2214 LOUVIERS
2215 LOVEJOY
2216 LOVELACE
2217 LOVELAND
2218 LOVELL
2219 LOVELOCK
2220 LOVEWELL
2221 LOVLINE
2222 LOWELL
2223 LOWERCREEK
2224 LOWLEIN
2225 LOWNDES
2226 LOWRY
2227 LOWS
2228 LOWVILLE
2229 LOX
2230 LOXLEY
2231 LOYAL
2232 LOYALTON
2233 LOYSVILLE
2234 LOZA
2235 LOZANO
2236 LOZIER
2237 LUALUALEI
2238 LUANA
2239 LUBBOCK
2240 LUBRECHT
2241 LUCAS
2242 LUCEDALE
2243 LUCERNE
2244 LUCERO
2245 LUCIEN
2246 LUCILE
2247 LUCKENBACH
2248 LUCKIAMUTE
2249 LUCKY
2250 LUCKY STAR
2251 LUCKYRICH
2252 LUCY
2253 LUD
2254 LUDDEN
2255 LUDINGTON
2256 LUDLOW
2257 LUEDERS
2258 LUFKIN
2259 LUGERT
2260 LUHON
2261 LUKE
2262 LUKIN
2263 LULA
2264 LULING
2265 LULUDE
2266 LUMBEE
2267 LUMMER
2268 LUMMI
2269 LUNA
2270 LUNCH
2271 LUNDER
2272 LUNDS
2273 LUNDY
2274 LUNING
2275 LUNT
2276 LUPE
2277 LUPINTO
2278 LUPOYOMA
2279 LUPPINO
2280 LUPTON
2281 LURA
2282 LURAY
2283 LURNICK
2284 LUSETTI
2285 LUTE
2286 LUTH
2287 LUTHER
2288 LUTIE
2289 LUTON
2290 LUTTERLOH
2291 LUTZKE
2292 LUVERNE
2293 LUXOR
2294 LUZENA
2295 LYBROOK
2296 LYDA
2297 LYDICK
2298 LYERLY
2299 LYFORD
2300 LYKENS
2301 LYLES
2302 LYMAN
2303 LYMANSON
2304 LYME
2305 LYNCH
2306 LYNCHBURG
2307 LYNDEN
2308 LYNN HAVEN
2309 LYNNDYL
2310 LYNNE
2311 LYNNVILLE
2312 LYNNWOOD
2313 LYNX
2314 LYONMAN
2315 LYONS
2316 LYONSVILLE
2317 LYRA
2318 LYRE
2319 LYSTAIR
2320 LYTELL
2321 LYVILLE
2322 LYX
2323 MABANK
2324 MABEN
2325 MABI
2326 MABRAY
2327 MACAR
2328 MACAREENO
2329 MACE
2330 MACEDONIA
2331 MACFARLANE
2332 MACHETE
2333 MACHIAS
2334 MACHUELO
2335 MACK
2336 MACKEN
2337 MACKEY
2338 MACKINAC
2339 MACKSBURG
2340 MACMEAL
2341 MACOMB
2342 MACOMBER
2343 MACON
2344 MADALIN
2345 MADAWASKA
2346 MADDEN
2347 MADDOCK
2348 MADELIA
2349 MADELINE
2350 MADERA
2351 MADGE
2352 MADILL
2353 MADISON
2354 MADONNA
2355 MADRAK
2356 MADRAS
2357 MADRID
2358 MADRONE
2359 MADUREZ
2360 MAES
2361 MAGALLON
2362 MAGENS
2363 MAGGIE
2364 MAGGIN
2365 MAGHILLS
2366 MAGIC
2367 MAGINNIS
2368 MAGNA
2369 MAGNET
2370 MAGNOR
2371 MAGNUS
2372 MAGOTSU
2373 MAHALA
2374 MAHALASVILLE
2375 MAHANA
2376 MAHASKA
2377 MAHONING
2378 MAHOOSUC
2379 MAHTOMEDI
2380 MAHTOWA

2381 MAHUKONA
2382 MAIA
2383 MAIDEN
2384 MAILE
2385 MAINSTAY
2386 MAITLAND
2387 MAJADA
2388 MAKAALAE
2389 MAKAH
2390 MAKALAPA
2391 MAKAPILI
2392 MAKAWAO
2393 MAKAWELI
2394 MAKENA
2395 MAKI
2396 MAKIKI
2397 MAKLAK
2398 MAKOTI
2399 MAL
2400 MALA
2401 MALABAR
2402 MALABON
2403 MALACHY
2404 MALAGA
2405 MALAMA
2406 MALARGO
2407 MALAYA
2408 MALBIS
2409 MALCOLM
2410 MALDEN
2411 MALEZA
2412 MALHEUR
2413 MALIBU
2414 MALIN
2415 MALIN, SILTY
2416 MALJAMAR
2417 MALLORY
2418 MALM
2419 MALMESA
2420 MALO
2421 MALOTERRE
2422 MALOY
2423 MALPAIS
2424 MALSTROM
2425 MALVERN
2426 MAMALA
2427 MAMOU
2428 MANAHAA
2429 MANAHAWKIN
2430 MANANA
2431 MANARD
2432 MANASSA
2433 MANASSAS
2434 MANASTASH
2435 MANATEE
2436 MANAWA
2437 MANBURN
2438 MANCELONA
2439 MANCHESTER
2440 MANDAN
2441 MANDARIN
2442 MANDERFIELD
2443 MANDEVILLE
2444 MANET
2445 MANFRED
2446 MANGUM
2447 MANHATTAN
2448 MANHEIM
2449 MANI
2450 MANILA
2451 MANISTEE
2452 MANITA
2453 MANITOWISH
2454 MANLEY
2455 MANLIUS
2456 MANN
2457 MANNING
2458 MANOGUE
2459 MANOR
2460 MANSELO
2461 MANSFIELD
2462 MANSIC
2463 MANSKER
2464 MANTACHIE
2465 MANTECA
2466 MANTEO
2467 MANTER
2468 MANU
2469 MANVEL
2470 MANZANITA
2471 MANZANO
2472 MANZANOLA
2473 MAPLE MOUNTAIN
2474 MAPLETON
2475 MARAGLADE
2476 MARAGUEZ
2477 MARASCHINO
2478 MARATHON
2479 MARBLE
2480 MARBLEMOUNT
2481 MARCELINAS
2482 MARCELLON
2483 MARCETTA
2484 MARCIAL
2485 MARCOLA
2486 MARCONI
2487 MARCOTT
2488 MARCOU
2489 MARCUM
2490 MARCUS
2491 MARCUSE
2492 MARCY
2493 MARDIN
2494 MARENGO
2495 MARESUA
2496 MARGATE
2497 MARGERUM
2498 MARGO
2499 MARIA
2500 MARIANA
2501 MARIAS
2502 MARIAVILLE
2503 MARICAO
2504 MARICOPA
2505 MARIETTA
2506 MARILLA
2507 MARIMEL
2508 MARINA
2509 MARINE
2510 MARION
2511 MARIPO
2512 MARIPOSA
2513 MARISCAL
2514 MARISSA
2515 MARKES
2516 MARKESAN
2517 MARKET
2518 MARKEY
2519 MARKHAM
2520 MARKLAND
2521 MARKLEPASS
2522 MARKTON
2523 MARLA
2524 MARLAKE
2525 MARLBORO
2526 MARLEAN
2527 MARLETTE
2528 MARLOW
2529 MARLTON
2530 MARMARTH
2531 MARNA
2532 MAROSA
2533 MARPA
2534 MARQUETTE
2535 MARQUEZ
2536 MARR
2537 MARRIOT
2538 MARRIOTT
2539 MARSDEN
2540 MARSEILLES
2541 MARSELL
2542 MARSHALL
2543 MARSHAN
2544 MARSHDALE
2545 MARSHFIELD
2546 MARSING
2547 MART
2548 MARTEL
2549 MARTELLA
2550 MARTIN
2551 MARTIN PENA
2552 MARTINECK
2553 MARTINEZ
2554 MARTINI
2555 MARTINSBURG
2556 MARTINSDALE
2557 MARTINSON
2558 MARTINSVILLE
2559 MARTINTON
2560 MARTIS
2561 MARTISCO
2562 MARTY
2563 MARVAN
2564 MARVELL
2565 MARVIN
2566 MARVYN
2567 MARY
2568 MARYSLAND

SOIL04.DAT

1 MASADA
2 MASARDIS
3 MASARYK
4 MASCAMP
5 MASCARENAS
6 MASCHETAH
7 MASCOTTE
8 MASET
9 MASHAM
10 MASHEL
11 MASHULAVILLE
12 MASKELL
13 MASON
14 MASONFORT
15 MASSACK
16 MASSANETTA
17 MASSBACH
18 MASSENA
19 MASSIE
20 MASTERSON
21 MATAGORDA
22 MATAMOROS
23 MATANUSKA
24 MATANZAS
25 MATAPEAKE
26 MATAWAN
27 MATCHER
28 MATFIELD
29 MATGO
30 MATHENY
31 MATHERS
32 MATHERTON
33 MATHESON
34 MATHIAS
35 MATHIS
36 MATHISTON
37 MATHON
38 MATLACHA
39 MATOY
40 MATTAMUSKEET
41 MATTAPEX
42 MATTAPONI
43 MATUNUCK
44 MAU
45 MAUDE
46 MAUGHAN
47 MAUKEY
48 MAULDIN
49 MAUMEE
50 MAUNABO
51 MAUPIN
52 MAUREPAS
53 MAURICE
54 MAURY
55 MAUSER
56 MAUVAIS
57 MAVCO
58 MAVERICK
59 MAVIE
60 MAWAE
61 MAWER
62 MAX
63 MAXCREEK
64 MAXEY
65 MAXFIELD
66 MAXTON
67 MAXVILLE
68 MAXWELL
69 MAY
70 MAY DAY
71 MAYACAMA
72 MAYBERRY
73 MAYBID
74 MAYDOL
75 MAYER
76 MAYES
77 MAYFIELD
78 MAYFLOWER
79 MAYGER
80 MAYHEW
81 MAYMEAD
82 MAYMEN
83 MAYNARD LAKE
84 MAYO
85 MAYODAN
86 MAYOWORTH
87 MAYOWORTH, DEEP
88 MAYQUEEN
89 MAYSDORF
90 MAYTOWN
91 MAYVILLE
92 MAYWOOD
93 MAZASKA
94 MAZUMA
95 MC BETH
96 MC CARRAN
97 MC CORT
98 MCAFEE
99 MCALLEN
100 MCALLISTER
101 MCALPIN
102 MCBEE
103 MCBIGGAM
104 MCBRIDE
105 MCCAFFERY
106 MCCAIN
107 MCCALEB
108 MCCALL
109 MCCALLY
110 MCCAMMON
111 MCCANN
112 MCCAREY
113 MCCARRAN
114 MCCARTHY
115 MCCASH
116 MCCLAVE
117 MCCLEARY
118 MCCLELLAN
119 MCCLOUD
120 MCCLURE
121 MCCOIN
122 MCCOLL
123 MCCOLLUM
124 MCCONNEL
125 MCCOOK
126 MCCORT
127 MCCOY
128 MCCOYSBURG
129 MCCREE
130 MCCRORY
131 MCCROSKET
132 MCCULLOUGH
133 MCCULLY
134 MCCUMBER
135 MCCUNE
136 MCCURDY
137 MCCUTCHEN
138 MCDANIEL
139 MCDOLE
140 MCDONALD
141 MCDONALDSVILLE
142 MCDUFF
143 MCELMO
144 MCELROY
145 MCEWEN
146 MCFADDEN
147 MCFAIN
148 MCFARLAND
149 MCFAUL
150 MCGAFFEY
151 MCGARR
152 MCGARY
153 MCGEHEE
154 MCGILVERY
155 MCGINNIS
156 MCGINTY
157 MCGIRK
158 MCGOWAN
159 MCGRATH
160 MCGREW
161 MCGUFFEY
162 MCGUIRE
163 MCHENRY
164 MCILWAINE
165 MCINTOSH
166 MCINTYRE
167 MCKAMIE
168 MCKAY
169 MCKENNA
170 MCKENZIE
171 MCKINLEY
172 MCKINNEY
173 MCKNIGHT
174 MCLAIN
175 MCLAURIN
176 MCLEOD
177 MCLOUGHLIN
178 MCMEEN
179 MCMILLE
180 MCMULLIN
181 MCMURDIE
182 MCMURRAY
183 MCNARY
184 MCNEAL
185 MCNULL
186 MCNULTY
187 MCPAUL
188 MCPHIE
189 MCQUARRIE
190 MCQUEEN
191 MCRAE
192 MCRAVEN
193 MCVEGAS
194 MCVICKERS
195 MEAD
196 MEADIN
197 MEADLAND
198 MEADOWCREEK
199 MEADOWLAKE
200 MEADOWVILLE
201 MECAN
202 MECHANICSBURG
203 MECKESVILLE
204 MECKLENBURG
205 MECOSTA
206 MEDA
207 MEDANO
208 MEDARY
209 MEDBURN
210 MEDCO
211 MEDFORD
212 MEDFRA
213 MEDICINE
214 MEDLEY
215 MEDLIN
216 MEDOMAK
217 MEDORA
218 MEDWAY
219 MEEGERO
220 MEEHAN
221 MEEKS
222 MEETEETSE
223 MEGALOS
224 MEGGETT
225 MEGONOT
226 MEGUIN
227 MEHLHORN
228 MEIKLE
229 MEISS
230 MELAKWA
231 MELAND
232 MELBOURNE
233 MELBY
234 MELD
235 MELDER
236 MELGA
237 MELHOMES
238 MELITA
239 MELLENTHIN
240 MELLOR
241 MELLOTT
242 MELOCHE
243 MELOLAND
244 MELON
245 MELROSE
246 MELTON
247 MELVILLE
248 MELVIN
249 MEMALOOSE
250 MEMPHIS
251 MENAHGA
252 MENARD
253 MENASHA
254 MENBO
255 MENDEBOURE
256 MENDENHALL
257 MENDI
258 MENDOCINO
259 MENDON
260 MENDOTA
261 MENEFEE
262 MENFRO
263 MENLO
264 MENO
265 MENOKEN
266 MENOMINEE
267 MENTO
268 MENTOR
269 MENZEL
270 MEQUON
271 MER ROUGE
272 MERCED
273 MERCEDES
274 MERCER
275 MERCEY
276 MERDEN
277 MEREDITH
278 MERETA
279 MERGEL
280 MERIDIAN

281 MERINO
282 MERKEL
283 MERLIN
284 MERMILL
285 MERNA
286 MEROS
287 MERRICK
288 MERRILL
289 MERRILLAN
290 MERRIMAC
291 MERRITT
292 MERSHON
293 MERTON
294 MERTZ
295 MERWIN
296 MESA
297 MESABA
298 MESCAL
299 MESCALERO
300 MESEI
301 MESPUN
302 MESSER
303 MET
304 METAMORA
305 METCALF
306 METEA
307 METH
308 METIGOSHE
309 METOLIUS
310 METRE
311 METZ
312 MEXICO
313 MEXISPRING
314 MEYSTRE
315 MHOON
316 MIAMI
317 MIAMIAN
318 MICANOPY
319 MICCO
320 MICHELSON
321 MICHIGAMME
322 MICKEY
323 MIDAS
324 MIDCO
325 MIDDLE
326 MIDDLEBURY
327 MIDDLETOWN
328 MIDDLEWOOD
329 MIDELIGHT
330 MIDESSA
331 MIDFORK
332 MIDLAND
333 MIDMONT
334 MIDNIGHT
335 MIDO
336 MIDRAW
337 MIDVALE
338 MIDWAY
339 MIERHILL
340 MIESEN
341 MIFFLIN
342 MIGUEL
343 MIKE
344 MIKESELL
345 MIKIM
346 MIKKALO
347 MILACA
348 MILAN
349 MILBURY
350 MILBY
351 MILCAN
352 MILES
353 MILFORD
354 MILHAM
355 MILITARY
356 MILL HOLLOW
357 MILLADORE
358 MILLBORO
359 MILLBROOK
360 MILLBURNE
361 MILLER
362 MILLERLAKE
363 MILLERLUX
364 MILLERTON
365 MILLERVILLE
366 MILLETT
367 MILLGROVE
368 MILLHEIM
369 MILLHOPPER
370 MILLICH
371 MILLICOMA
372 MILLIGAN
373 MILLING
374 MILLINGTON
375 MILLIS
376 MILLPAW
377 MILLPOT
378 MILLROCK
379 MILLSAP
380 MILLSDALE
381 MILLSHOLM
382 MILLSITE
383 MILLVILLE
384 MILLWOOD
385 MILNER
386 MILOK
387 MILPITAS
388 MILREN
389 MILTON
390 MIMBRES
391 MIMOSA
392 MINA
393 MINALOOSA
394 MINAT
395 MINATARE
396 MINCHEY
397 MINCO
398 MINDEN
399 MINE
400 MINEOLA
401 MINER
402 MINERAL
403 MINERAL MOUNTAIN
404 MINESINGER
405 MINGO
406 MINGUS
407 MINIDOKA
408 MINKLER
409 MINLITH
410 MINNEHA
411 MINNEISKA
412 MINNEOPA
413 MINNEOSA
414 MINNEQUA
415 MINNETONKA
416 MINNEWAUKAN
417 MINNIECE
418 MINNIEPEAK
419 MINNIMAUD
420 MINNITH
421 MINNYE
422 MINOA
423 MINOCQUA
424 MINTER
425 MINTO
426 MINU
427 MINVALE
428 MINVENO
429 MINWELLS
430 MION
431 MIPPON
432 MIRABAL
433 MIRACLE
434 MIRAGE
435 MIRAND
436 MIRANDA
437 MIRES
438 MIRKWOOD
439 MIRROR
440 MIRROR LAKE
441 MISAD
442 MISENHEIMER
443 MISHAK
444 MISSION
445 MISSISQUOI
446 MISSLER
447 MISSOULA
448 MITCH
449 MITCHELL
450 MITIWANGA
451 MITRE
452 MITTEN
453 MIVIDA
454 MIZEL
455 MIZPAH
456 MOAB
457 MOAG
458 MOANO
459 MOAPA
460 MOAULA
461 MOBATE
462 MOBEETIE
463 MOBERG
464 MOBIL
465 MOBL
466 MOBRIDGE
467 MOCA
468 MOCAREY
469 MOCHO
470 MOCKLER
471 MOCMONT
472 MOCTILEME
473 MODA
474 MODALE
475 MODENA
476 MODESTO
477 MODJESKA
478 MODKIN
479 MODOC
480 MODYON
481 MOE
482 MOEN
483 MOENKOPIE
484 MOEPITZ
485 MOFFAT
486 MOGG
487 MOGLIA
488 MOGOLLON
489 MOGOTE
490 MOHALL
491 MOHAVE
492 MOHAWK
493 MOIESE
494 MOINGONA
495 MOKELUMNE
496 MOKENA
497 MOKIAK
498 MOKINS
499 MOKO
500 MOKULEIA
501 MOLALLA
502 MOLAND
503 MOLAS
504 MOLCAL
505 MOLENA
506 MOLION
507 MOLLIC HALAQUEPT
508 MOLLICY
509 MOLLMAN
510 MOLLVILLE
511 MOLLY
512 MOLOKAI
513 MOLSON
514 MOLYNEUX
515 MOMOLI
516 MONA
517 MONACAN
518 MONACHE
519 MONAD
520 MONADNOCK
521 MONAHANS
522 MONARDA
523 MONASTERIO
524 MONAVILLE
525 MONBUTTE
526 MONDAMIN
527 MONDEY
528 MONDOVI
529 MONEE
530 MONICO
531 MONIDA
532 MONIERCO
533 MONITEAU
534 MONITOR
535 MONJEAU
536 MONOGRAM
537 MONONA
538 MONONGAHELA
539 MONROE
540 MONROEVILLE
541 MONSERATE
542 MONSON
543 MONTAGUE
544 MONTALTO
545 MONTARA
546 MONTAUK
547 MONTBORNE
548 MONTCALM
549 MONTE
550 MONTE CRISTO
551 MONTECITO
552 MONTEGRANDE
553 MONTELL
554 MONTELLO
555 MONTEOCHA
556 MONTEOLA
557 MONTEROSA
558 MONTESA
559 MONTEVALLO
560 MONTEZ
561 MONTGOMERY
562 MONTICELLO
563 MONTIETH
564 MONTLID
565 MONTMORENCI
566 MONTNEVA
567 MONTOSA
568 MONTOSO
569 MONTOUR
570 MONTOYA
571 MONTPELLIER
572 MONTROSS
573 MONTVALE
574 MONTVERDE
575 MONTWEL
576 MONUE
577 MOODY
578 MOOHOO
579 MOOLACK
580 MOONLIGHT

581 MOONSHINE
582 MOONSTONE
583 MOONVILLE
584 MOOREVILLE
585 MOOSE RIVER
586 MOOSELAKE
587 MOOSILAUKE
588 MOPANG
589 MOQUAH
590 MORA
591 MORADO
592 MORALES
593 MORAN
594 MORANCH
595 MORD
596 MOREAU
597 MOREHEAD
598 MOREHOUSE
599 MORELAND
600 MORENO
601 MORET
602 MOREY
603 MORGALA
604 MORGANFIELD
605 MORIARTY
606 MORICAL
607 MORLEY
608 MORLING
609 MORMAN MESA
610 MORMON MESA
611 MOROCCO
612 MORONI
613 MOROP
614 MORPH
615 MORRILL
616 MORRIS
617 MORRISON
618 MORRISTOWN
619 MORROW
620 MORSE
621 MORSET
622 MORTENSON
623 MORTON
624 MORVAL
625 MOSBY
626 MOSCA
627 MOSCOW
628 MOSEL
629 MOSHANNON
630 MOSHEIM
631 MOSHER
632 MOSHERVILLE
633 MOSHUP
634 MOSIDA
635 MOSINEE
636 MOSLANDER
637 MOSO
638 MOSQUET
639 MOSROC
640 MOSSYROCK
641 MOSWELL
642 MOTA
643 MOTEN
644 MOTLEY
645 MOTOQUA
646 MOTT
647 MOTTLAND
648 MOTTO
649 MOTTSVILLE
650 MOULTON
651 MOULTRIE
652 MOUND
653 MOUNDPRAIRIE
654 MOUNDVILLE
655 MOUNT HOME
656 MOUNT LUCAS
657 MOUNTAINBOY
658 MOUNTAINBURG
659 MOUNTAINEER
660 MOUNTAINVIEW
661 MOUNTAINVILLE
662 MOUNTVIEW
663 MOVILLE
664 MOWATA
665 MOWEBA
666 MOWER
667 MOWICH
668 MOXEE
669 MOYERS
670 MOYERSON
671 MOYINA
672 MT. AIRY
673 MT. CARROLL
674 MT. VERNON
675 MUCARA
676 MUCK
677 MUCKALEE
678 MUD SPRINGS
679 MUDRAY
680 MUDSOCK
681 MUES
682 MUFF
683 MUGGINS
684 MUGHOUSE
685 MUGHUT
686 MUIR
687 MUIRKIRK
688 MUKILTEO
689 MULAT
690 MULDOON
691 MULDROW
692 MULETT
693 MULGON
694 MULHALL
695 MULHOP
696 MULKEY
697 MULLICA
698 MULLIG
699 MULLINS
700 MULLYON
701 MULSHOE
702 MULSTAY
703 MULT
704 MULTEY
705 MULTNOMAH
706 MULTORPOR
707 MUNDAL
708 MUNDELEIN
709 MUNDEN
710 MUNDOS
711 MUNDT
712 MUNI
713 MUNISING
714 MUNJOR
715 MUNK
716 MUNSET
717 MUNSON
718 MUNUSCONG
719 MURAD
720 MURDO
721 MURDOCK
722 MUREN
723 MURNEN
724 MUROC
725 MURPHY
726 MURRIETA
727 MURRILL
728 MURVILLE
729 MUSCATINE
730 MUSE
731 MUSELLA
732 MUSICK
733 MUSINA
734 MUSINIA
735 MUSKEGO
736 MUSKELLUNGE
737 MUSKINGUM
738 MUSKOGEE
739 MUSQUIZ
740 MUSSEL
741 MUSSELSHELL
742 MUSSEY
743 MUSTANG
744 MUTNALA
745 MUZZLER
746 MYAKKA
747 MYATT
748 MYERS
749 MYERSVILLE
750 MYFORD
751 MYLREA
752 MYOMA
753 MYRICK
754 MYRTLE
755 MYSTEN
756 MYSTIC
757 MYTON
758 NAALEHU
759 NABESNA
760 NACHES
761 NACHUSA
762 NACIMIENTO
763 NACLINA
764 NACOGDOCHES
765 NADA
766 NADEAU
767 NADINA
768 NADRA
769 NAEGELIN
770 NAFF
771 NAGITSY
772 NAGLE
773 NAGROM
774 NAHATCHE
775 NAHMA
776 NAHON
777 NAHRUB
778 NAHUNTA
779 NAIWA
780 NAKAI
781 NAKARNA
782 NAKNEK
783 NALAKI
784 NALDO
785 NALL
786 NAMBE
787 NAMEOKI
788 NAMON
789 NAMUR
790 NANAMKIN
791 NANCY
792 NANIAK
793 NANKIN
794 NANNY
795 NANNYTON
796 NANSEMOND
797 NANSENE
798 NANSEPSEP
799 NANTUCKET
800 NANUM
801 NAPA
802 NAPIER
803 NAPLENE
804 NAPOLEON
805 NAPPANEE
806 NAPTOWNE
807 NARANJITO
808 NARANJO
809 NARCISSE
810 NARCOOSSEE
811 NARD
812 NAREL
813 NARGAR
814 NARK
815 NARLON
816 NARON
817 NARRAGANSETT
818 NARROWS
819 NARTA
820 NARU
821 NASER
822 NASH
823 NASHMEAD
824 NASHOBA
825 NASHVILLE
826 NASHWAUK
827 NASON
828 NASS
829 NASSAU
830 NASSET
831 NATAL
832 NATAL VARIANT
833 NATCHEZ
834 NATCHITOCHES
835 NATHROP
836 NATI
837 NATIONAL
838 NATKIM
839 NATROY
840 NATURITA
841 NAUMBURG
842 NAUVOO
843 NAVACA
844 NAVAJO
845 NAVAN
846 NAVIDAD
847 NAVINA
848 NAVO
849 NAWNEY
850 NAWT
851 NAXING
852 NAYE
853 NAYPED
854 NAYRIB
855 NAZ
856 NAZATON
857 NEABSCO
858 NEBAGO
859 NEBEKER
860 NEBGEN
861 NEBISH
862 NEBONA
863 NECANICUM
864 NECESSITY
865 NECHE
866 NECONDA
867 NECTAR
868 NEDA
869 NEDERLAND
870 NEEDLE
871 NEEDLE PEAK
872 NEEDLEYE
873 NEEDMORE
874 NEELEY
875 NEEN
876 NEENAH
877 NEER
878 NEESES
879 NEESOPAH
880 NEETO

881 NEFF
882 NEGLEY
883 NEHALEM
884 NEHAR
885 NEIBER
886 NEICE
887 NEILTON
888 NEISSENBERG
889 NEKIA
890 NEKKEN
891 NEKOMA
892 NELDORE
893 NELLA
894 NELLIS
895 NELMAN
896 NELSCOTT
897 NELSON
898 NEMADJI
899 NEMAH
900 NEMICO
901 NEMOTE
902 NEMOURS
903 NENANA
904 NENNO
905 NEOTOMA
906 NEPALTO
907 NEPESTA
908 NEPHI
909 NEPONSET
910 NEPPEL
911 NEPTUNE
912 NERESON
913 NESBITT
914 NESDA
915 NESHAMINY
916 NESHOBA
917 NESIKA
918 NESKAHI
919 NESKOWIN
920 NESO
921 NESPELEM
922 NESS
923 NESSEL
924 NESTER
925 NESTORIA
926 NESTUCCA
927 NET
928 NETARTS
929 NETCONG
930 NETO
931 NETRAC
932 NETTLES
933 NETTLETON
934 NEUBERT
935 NEUNS
936 NEURALIA
937 NEUSKE
938 NEVADOR
939 NEVARC
940 NEVAT
941 NEVEE
942 NEVILLE
943 NEVIN
944 NEVINE
945 NEVKA
946 NEVOYER
947 NEVTAH
948 NEVU
949 NEW CAMBRIA
950 NEWALBIN
951 NEWALLA
952 NEWANNA
953 NEWARK
954 NEWAUKUM
955 NEWAYGO
956 NEWBELL
957 NEWBERG
958 NEWBERN
959 NEWBERRY
960 NEWBON
961 NEWCO
962 NEWCOMB
963 NEWDALE
964 NEWELL
965 NEWELLTON
966 NEWFIELDS
967 NEWFLAT
968 NEWFORK
969 NEWFOUND
970 NEWGLARUS
971 NEWHAN
972 NEWHOUSE
973 NEWKIRK
974 NEWLANDS
975 NEWLIN
976 NEWNAN
977 NEWNATA
978 NEWPARTY
979 NEWPASS
980 NEWPORT
981 NEWRY
982 NEWSKAH
983 NEWSON
984 NEWSROCK
985 NEWSTEAD
986 NEWTON
987 NEWTONIA
988 NEWTOWN
989 NEWULM
990 NEWVIENNA
991 NEWVILLE
992 NEYGAT
993 NEZ PERCE
994 NGARDMAU
995 NGARDOK
996 NGATPANG
997 NGEDEBUS
998 NGERSUUL
999 NGERUNGOR
1000 NIAGARA
1001 NIARADA
1002 NIART
1003 NIBLEY
1004 NIBSON
1005 NICANOR
1006 NICASIO
1007 NICHOLFLAT
1008 NICHOLIA
1009 NICHOLS
1010 NICHOLSON
1011 NICHOLVILLE
1012 NICKEL
1013 NICKIN
1014 NICODEMUS
1015 NICOLLET
1016 NIDO
1017 NIELSEN
1018 NIGHTHAWK
1019 NIHILL
1020 NIKEY
1021 NIKFUL
1022 NIKISHKA
1023 NIKLASON
1024 NIKOLAI
1025 NILAND
1026 NILER
1027 NILRAP
1028 NIMBRO
1029 NIMERICK
1030 NIMMO
1031 NIMROD
1032 NIMS
1033 NIMUE
1034 NINCH
1035 NINEKAR
1036 NINEMILE
1037 NINEVEH
1038 NINIGRET
1039 NIOBELL
1040 NIOTA
1041 NIOTAZE
1042 NIPE
1043 NIPINTUCK
1044 NIPPT
1045 NIPSUM
1046 NIRA
1047 NISENE
1048 NISHNA
1049 NISHON
1050 NISQUALLY
1051 NISULA
1052 NITTAW
1053 NIU
1054 NIULII
1055 NIWANA
1056 NIWOT
1057 NIXA
1058 NIXON
1059 NIXONTON
1060 NIZINA
1061 NOARK
1062 NOBE
1063 NOBLE
1064 NOBLETON
1065 NOBSCOT
1066 NOCKEN
1067 NODAWAY
1068 NODEN
1069 NODINE
1070 NOELKE
1071 NOGAL
1072 NOHILI
1073 NOKASIPPI
1074 NOKAY
1075 NOKHU
1076 NOLAM
1077 NOLICHUCKY
1078 NOLIN
1079 NOLO
1080 NOLTEN
1081 NOMARA
1082 NOME
1083 NOMIE
1084 NONDALTON
1085 NONOPAHU
1086 NONPAREIL
1087 NOOK
1088 NOOKACHAMPS
1089 NOOKSACK
1090 NOONAN
1091 NORA
1092 NORAD
1093 NORBERT
1094 NORBORNE
1095 NORCAN
1096 NORD
1097 NORDBY
1098 NORDEN
1099 NORDIC
1100 NORDICOL
1101 NORDNESS
1102 NORFOLK
1103 NORFORK
1104 NORGE
1105 NORGO
1106 NORKA
1107 NORKOOL
1108 NORLAND
1109 NORMA
1110 NORMANGEE
1111 NORMANIA
1112 NOROB
1113 NORREST
1114 NORRIE
1115 NORRIS
1116 NORRISTON
1117 NORTE
1118 NORTEZ
1119 NORTH POWDER
1120 NORTHBORO
1121 NORTHCASTLE
1122 NORTHCOTE
1123 NORTHDALE
1124 NORTHFIELD
1125 NORTHMORE
1126 NORTHWATER
1127 NORTHWOOD
1128 NORTON
1129 NORTONVILLE
1130 NORWELL
1131 NORWICH
1132 NORWOOD
1133 NOSRAC
1134 NOTAL
1135 NOTCHER
1136 NOTI
1137 NOTSPIER
1138 NOTTAWA
1139 NOTTER
1140 NOTUS
1141 NOUQUE
1142 NOVACAN
1143 NOVARK
1144 NOVARY
1145 NOVATO
1146 NOVINA
1147 NOWATA
1148 NOWEN
1149 NOYER
1150 NOYES
1151 NOYO
1152 NOYSON
1153 NUBY
1154 NUC
1155 NUCKOLLS
1156 NUCLA
1157 NUECES
1158 NUEVA
1159 NUFF
1160 NUGENT
1161 NUKRUM
1162 NULEY
1163 NULLIGAM
1164 NUMA
1165 NUNDA
1166 NUNEMAKER
1167 NUNICA
1168 NUNN
1169 NUNNSTON
1170 NURKEY
1171 NUSMAG
1172 NUSS
1173 NUTIVOLI
1174 NUTLEY
1175 NUTRAS
1176 NUTRIOSO
1177 NUVALDE
1178 NYALA
1179 NYCON
1180 NYE

1181 NYJACK
1182 NYMORE
1183 NYSERVA
1184 NYSSA
1185 NYSSATON
1186 O'BRIEN
1187 O'NEILL
1188 OAHE
1189 OAK GLEN
1190 OAK GROVE
1191 OAKALLA
1192 OAKBORO
1193 OAKDEN
1194 OAKES
1195 OAKHILL SERIES
1196 OAKHURST
1197 OAKLAND
1198 OAKLET
1199 OAKLIMETER
1200 OAKVILLE
1201 OAKWOOD
1202 OANAPUKA
1203 OATLANDS
1204 OATMAN
1205 OATUU
1206 OBAN
1207 OBANION
1208 OBARO
1209 OBEN
1210 OBISPO
1211 OBRAST
1212 OBRAY
1213 OBURN
1214 OCALA
1215 OCAMBEE
1216 OCANA
1217 OCCOQUAN
1218 OCCUM
1219 OCEANET
1220 OCEANO
1221 OCHEYEDAN
1222 OCHLOCKONEE
1223 OCHO
1224 OCHOCO
1225 OCIE
1226 OCILLA
1227 OCKLEY
1228 OCOEE
1229 OCONEE
1230 OCONTO
1231 OCOSTA
1232 OCQUEOC
1233 OCRAIG
1234 OCTAGON
1235 OCTAVIA
1236 ODAS
1237 ODELL
1238 ODEM
1239 ODENSON
1240 ODERMOTT
1241 ODESSA
1242 ODIN
1243 ODNE
1244 ODO
1245 OELOP
1246 OEST
1247 OESTERLE
1248 OFU
1249 OGARTY
1250 OGEECHEE
1251 OGEMAW
1252 OGILVIE
1253 OGLALA
1254 OGLE
1255 OGLESBY
1256 OGRAL
1257 OHACO
1258 OHANA
1259 OHIA
1260 OHOP
1261 OIDEM
1262 OIL-WASTE LAND
1263 OJATA
1264 OJIBWAY
1265 OJITO
1266 OKANOGAN
1267 OKATON
1268 OKAW
1269 OKAY
1270 OKEE
1271 OKEECHOBEE
1272 OKEELANTA
1273 OKEETEE
1274 OKEMAH
1275 OKIOTA
1276 OKLARED
1277 OKLAWAHA
1278 OKO
1279 OKOBOJI
1280 OKOLONA
1281 OKREEK
1282 OKTIBBEHA
1283 OLA
1284 OLAA
1285 OLAC
1286 OLAND
1287 OLANTA
1288 OLASHES
1289 OLBUT
1290 OLD CAMP
1291 OLDHAM
1292 OLDS
1293 OLDSFERRY
1294 OLDSMAR
1295 OLELO
1296 OLENO
1297 OLENTANGY
1298 OLEQUA
1299 OLETE
1300 OLEX
1301 OLGA
1302 OLI
1303 OLIAGA
1304 OLIN
1305 OLINDA
1306 OLIPHANT
1307 OLIVENHAIN
1308 OLIVIER
1309 OLJETO
1310 OLLEI
1311 OLLIE
1312 OLMITO
1313 OLMITZ
1314 OLMOS
1315 OLMSTED
1316 OLNES
1317 OLNEY
1318 OLOAVA
1319 OLOKUI
1320 OLOMOUNT
1321 OLOMPALI
1322 OLOT
1323 OLOTANIA
1324 OLPE
1325 OLSON
1326 OLTON
1327 OLUSTEE
1328 OLYIC
1329 OLYMPIC
1330 OMADI
1331 OMEGA
1332 OMENA
1333 OMIO
1334 OMNI
1335 OMRO
1336 OMSTOTT
1337 OMULGA
1338 ONA
1339 ONAMIA
1340 ONAQUI
1341 ONARGA
1342 ONASON
1343 ONAWA
1344 ONAWAY
1345 ONDAWA
1346 ONECO
1347 ONEIL
1348 ONITA
1349 ONITE
1350 ONOTA
1351 ONSLOW
1352 ONTARIO
1353 ONTKO
1354 ONTONAGON
1355 ONYX
1356 OOKALA
1357 OOSEN
1358 OPAL
1359 OPELIKA
1360 OPEQUON
1361 OPHIR
1362 OPIHIKAO
1363 OPLIN
1364 OPPIO
1365 OQUAGA
1366 ORA
1367 ORACLE
1368 ORAGRAN
1369 ORAID
1370 ORAN
1371 ORANGE
1372 ORANGEBURG
1373 ORCAP
1374 ORCAS
1375 ORCHARD
1376 ORCKY
1377 ORD
1378 ORDNANCE
1379 ORDWAY
1380 OREJAS
1381 ORELIA
1382 ORELIA,CLAYEY SU
1383 ORELLA
1384 ORFORD
1385 ORICTO
1386 ORIDIA
1387 ORIF
1388 ORIGO
1389 ORINOCO
1390 ORIO
1391 ORION
1392 ORITA
1393 ORIZABA
1394 ORLA
1395 ORLAND
1396 ORLANDO
1397 ORLIE
1398 ORMAS
1399 ORMISTON
1400 ORMSBY
1401 ORNBAUN
1402 ORO FINO
1403 ORO GRANDE
1404 OROGNEN
1405 ORONOCO
1406 OROVADA
1407 ORPARK
1408 ORPHA
1409 ORPHANT
1410 ORR
1411 ORRUB
1412 ORRVILLE
1413 ORSA
1414 ORSET
1415 ORSINO
1416 ORTEGA
1417 ORTELLO
1418 ORTHENTS
1419 ORTHIDS
1420 ORTING
1421 ORTIZ
1422 ORTON
1423 ORWET
1424 ORWIG
1425 ORWOOD
1426 OSAGE
1427 OSAKIS
1428 OSBORN
1429 OSCAR
1430 OSCURA
1431 OSGOOD
1432 OSHA
1433 OSHAWA
1434 OSHKOSH
1435 OSHONE
1436 OSHTEMO
1437 OSIER
1438 OSITO
1439 OSKA
1440 OSMUND
1441 OSO
1442 OSOBB
1443 OSOLL
1444 OSORIDGE
1445 OSOTE
1446 OSSIAN
1447 OSSIPEE
1448 OST
1449 OSTLER
1450 OSTRANDER
1451 OSWALD
1452 OTANYA
1453 OTEEN
1454 OTERO
1455 OTHELLO
1456 OTISCO
1457 OTISVILLE
1458 OTLEY
1459 OTOMO
1460 OTOOLE
1461 OTTER
1462 OTTERHOLT
1463 OTTERSON
1464 OTTOKEE
1465 OTTOSEN
1466 OTWAY
1467 OTWELL
1468 OTWIN
1469 OUACHITA
1470 OUARD
1471 OULA
1472 OUPICO
1473 OURAY
1474 OUSLEY
1475 OUTLET
1476 OUTLOOK
1477 OVALL
1478 OVAN
1479 OVERGAARD
1480 OVERLAND

1481 OVERLY
1482 OVERTON
1483 OVIATT
1484 OVID
1485 OVINA
1486 OWEGO
1487 OWEN CREEK
1488 OWENS
1489 OWHI
1490 OWINZA
1491 OWLCAN
1492 OWOSSO
1493 OWSEL
1494 OWYHEE
1495 OXALIS
1496 OXBOW
1497 OXCOREL
1498 OXERINE
1499 OXFORD
1500 OXHEAD
1501 OXLEY
1502 OXWALL
1503 OYHUT
1504 OZAMIS
1505 OZAN
1506 OZAUKEE
1507 OZETTE
1508 OZIAS
1509 PAAIKI
1510 PAALOA
1511 PAAUHAU
1512 PABLO
1513 PACHAPPA
1514 PACHECO
1515 PACHIC CRYOBORO
1516 PACIFICO
1517 PACK
1518 PACKARD
1519 PACKER
1520 PACKHAM
1521 PACKTRAIL
1522 PACKWOOD
1523 PACO
1524 PACOLET
1525 PACTOLA
1526 PACTOLUS
1527 PADDOCK
1528 PADEN
1529 PADILLA
1530 PADINA
1531 PADRONES
1532 PADUCAH
1533 PADUS
1534 PAESL
1535 PAGEBROOK
1536 PAGET
1537 PAGODA
1538 PAGOSA
1539 PAHOKEE
1540 PAHRANAGAT
1541 PAHRANGE
1542 PAHREAH
1543 PAHROC
1544 PAHRUMP
1545 PAHSIMEROI
1546 PAIA
1547 PAICE
1548 PAINESVILLE
1549 PAINT
1550 PAISLEY
1551 PAIT
1552 PAJARA
1553 PAJARITO
1554 PAJARO
1555 PAJUELA
1556 PAKA
1557 PAKALA
1558 PAKINI
1559 PALAFOX
1560 PALANUSH
1561 PALAPALAI
1562 PALATINE
1563 PALAU
1564 PALAZZO
1565 PALEAQUULTS
1566 PALINOR
1567 PALISADE
1568 PALIX
1569 PALLS
1570 PALM BEACH
1571 PALMA
1572 PALMAREJO
1573 PALMAS ALTAS
1574 PALMER CANYON
1575 PALMERDALE
1576 PALMETTO
1577 PALMICH
1578 PALMS
1579 PALMYRA
1580 PALODURO
1581 PALOMARIN
1582 PALOMAS
1583 PALOMINO
1584 PALON
1585 PALOPINTO
1586 PALOS VERDES
1587 PALOUSE
1588 PALSGROVE
1589 PALUXY
1590 PAMLICO
1591 PAMOA
1592 PAMSDEL
1593 PAMUNKEY
1594 PANA
1595 PANAEWA
1596 PANAMA
1597 PANAMINT
1598 PANASOFFKEE
1599 PANCHERI
1600 PANDO
1601 PANDOAH
1602 PANDORA
1603 PANDURA
1604 PANE
1605 PANGBORN
1606 PANGUITCH
1607 PANIN
1608 PANIOGUE
1609 PANITCHEN
1610 PANKY
1611 PANMOD
1612 PANOCHE
1613 PANOLA
1614 PANORAMA
1615 PANSEY
1616 PANTANO
1617 PANTEGO
1618 PANTERA
1619 PANTHER
1620 PANTON
1621 PAOLA
1622 PAOLI
1623 PAPAA
1624 PAPAC
1625 PAPAGUA
1626 PAPAI
1627 PAPALOTE
1628 PAPINEAU
1629 PAPOOSE
1630 PARA
1631 PARACHUTE
1632 PARADISE
1633 PARANAT
1634 PARASOL
1635 PARCELAS
1636 PARCHIN
1637 PARDALOE
1638 PARDEE
1639 PARDEEVILLE
1640 PAREHAT
1641 PARENT
1642 PARIETTE
1643 PARISIAN
1644 PARKAY
1645 PARKDALE
1646 PARKE
1647 PARKER
1648 PARKFIELD
1649 PARKHILL
1650 PARKINSON
1651 PARKS
1652 PARKVIEW
1653 PARKVILLE
1654 PARKWOOD
1655 PARLEYS
1656 PARLIN
1657 PARLO
1658 PARMELE
1659 PARMELOW
1660 PARMENTER
1661 PARMLEE
1662 PARNELL
1663 PARQUAT
1664 PARR
1665 PARRAN
1666 PARRISH
1667 PARRITA
1668 PARSHALL
1669 PARSIPPANY
1670 PARSONS
1671 PARTLOW
1672 PARTOV
1673 PARTRI
1674 PASAGSHAK
1675 PASCO
1676 PASO SECO
1677 PASQUETTI
1678 PASQUOTANK
1679 PASS CANYON
1680 PASSAR
1681 PASSCREEK
1682 PASTERN
1683 PASTIK
1684 PASTORIUS
1685 PASTURA
1686 PATAHA
1687 PATCHIN
1688 PATE
1689 PATENT
1690 PATHEAD
1691 PATILLAS
1692 PATILO
1693 PATIO
1694 PATIT CREEK
1695 PATNA
1696 PATOS
1697 PATOUTVILLE
1698 PATRICIA
1699 PATRICK
1700 PATROLE
1701 PATTANI
1702 PATTEE
1703 PATTENBURG
1704 PATTER
1705 PATTERSON
1706 PATTON
1707 PAUL
1708 PAULDING
1709 PAULINA
1710 PAULSON
1711 PAULVILLE
1712 PAUMALU
1713 PAUNSAUGUNT
1714 PAUSANT
1715 PAUWELA
1716 PAVAIAI
1717 PAVANT
1718 PAVILLION
1719 PAVO
1720 PAVOHROO
1721 PAWCATUCK
1722 PAWHUSKA
1723 PAWLING
1724 PAWNEE
1725 PAXTON
1726 PAXVILLE
1727 PAYETTE
1728 PAYMASTER
1729 PAYNE
1730 PAYNECREEK
1731 PAYSON
1732 PAZAR
1733 PEACHAM
1734 PEARL
1735 PEARL HARBOR
1736 PEARSOLL
1737 PEASLEY
1738 PEASPEAR
1739 PEAVINE
1740 PEAWICK
1741 PEBBLEPOINT
1742 PECATONICA
1743 PECKHAM
1744 PECKISH
1745 PECOS
1746 PECTURE
1747 PEDEE
1748 PEDERNALES
1749 PEDIGO
1750 PEDLEFORD
1751 PEDOLI
1752 PEDRICK
1753 PEDRO
1754 PEEBLES
1755 PEEKO
1756 PEEL
1757 PEELER
1758 PEETZ
1759 PEEVER
1760 PEEVYWELL
1761 PEGLEG
1762 PEGLER
1763 PEGRAM
1764 PEKAY
1765 PEKIN
1766 PELAHATCHIE
1767 PELAN
1768 PELEE
1769 PELELIU
1770 PELHAM
1771 PELIC
1772 PELION
1773 PELKIE
1774 PELLA
1775 PELLEJAS
1776 PELLICER
1777 PELONCILLO
1778 PELTIER
1779 PEMBERTON
1780 PEMBROKE

1781 PEMENE
1782 PENA
1783 PENAPON
1784 PENASCO
1785 PENCE
1786 PEND OREILLE
1787 PEND ORIELLE
1788 PENDARVIS
1789 PENDEN
1790 PENDER
1791 PENDERGRASS
1792 PENDLETON
1793 PENDROY
1794 PENELAS
1795 PENGILLY
1796 PENGRA
1797 PENINSULA
1798 PENISTAJA
1799 PENITENTE
1800 PENLAW
1801 PENN
1802 PENNEKAMP
1803 PENNELL
1804 PENNICHUCK
1805 PENNSUCO
1806 PENNY
1807 PENO
1808 PENOYER
1809 PENROSE
1810 PENSORE
1811 PENTHOUSE
1812 PENTZ
1813 PENWELL
1814 PENWOOD
1815 PEOGA
1816 PEOH
1817 PEONE
1818 PEORIA
1819 PEOTONE
1820 PEPAL
1821 PEPOON
1822 PEPPER
1823 PEQUAMING
1824 PEQUEA
1825 PEQUOD
1826 PEQUOP
1827 PERAZZO
1828 PERCETON
1829 PERCHAS
1830 PERCILLA
1831 PERCIVAL
1832 PERCOUN
1833 PERCY
1834 PERDIN
1835 PERELLA
1836 PERHAM
1837 PERICO
1838 PERIDGE
1839 PERITSA
1840 PERKINS
1841 PERKS
1842 PERLA
1843 PERMA
1844 PERN
1845 PERNITAS
1846 PERNTY
1847 PERQUIMANS
1848 PERRIN
1849 PERRINE
1850 PERRINTON
1851 PERRY
1852 PERRYPARK
1853 PERRYVILLE
1854 PERSANTI
1855 PERSAYO
1856 PERSHING
1857 PERSIS
1858 PERT
1859 PERU
1860 PERVINA
1861 PERWICK
1862 PESCADERO
1863 PESCAR
1864 PESERO
1865 PESHASTIN
1866 PESHEKEE
1867 PESMO
1868 PESO
1869 PESOWYO
1870 PETACA
1871 PETAL
1872 PETAN
1873 PETEETNEET
1874 PETERMAN
1875 PETERS
1876 PETERSON
1877 PETESCREEK
1878 PETRIE
1879 PETROLIA
1880 PETROS
1881 PETSPRING
1882 PETTICOAT
1883 PETTIGREW
1884 PETTUS
1885 PETTY
1886 PEWAMO
1887 PEYTON
1888 PFEIFFER
1889 PHAGE
1890 PHALANX
1891 PHANTOM
1892 PHARO
1893 PHARR
1894 PHEBA
1895 PHEENEY
1896 PHELAN
1897 PHELPS
1898 PHERSON
1899 PHIFERSON
1900 PHILBON
1901 PHILDER
1902 PHILIPPA
1903 PHILIPSBURG
1904 PHILLIPS
1905 PHILO
1906 PHILOMATH
1907 PHING
1908 PHIPPS
1909 PHOEBE
1910 PHOENIX
1911 PHYS
1912 PIANKESHAW
1913 PIASA
1914 PIBLER
1915 PICABO
1916 PICACHO
1917 PICANTE
1918 PICAYUNE
1919 PICEANCE
1920 PICKAWAY
1921 PICKENS
1922 PICKETT
1923 PICKFORD
1924 PICKNEY
1925 PICKRELL
1926 PICKTON
1927 PICKUP
1928 PICKWICK
1929 PICO
1930 PICOSA
1931 PIDCOKE
1932 PIDINEEN
1933 PIE CREEK
1934 PIERIAN
1935 PIERPONT
1936 PIERRE
1937 PIERSONTE
1938 PIETOWN
1939 PIGTAIL
1940 PIIHONUA
1941 PIKE
1942 PIKEVILLE
1943 PILCHUCK
1944 PILEUP
1945 PILINE
1946 PILLIKEN
1947 PILLOT
1948 PILLSBURY
1949 PILOT PEAK
1950 PILOT ROCK
1951 PILTDOWN
1952 PILTZ
1953 PIMA
1954 PIMER
1955 PINAL
1956 PINALENO
1957 PINAMT
1958 PINATA
1959 PINAVETES
1960 PINBIT
1961 PINCHER
1962 PINCHOT
1963 PINCKNEY
1964 PINCONNING
1965 PINE FLAT
1966 PINEAL
1967 PINEBUTTE
1968 PINEDA
1969 PINEDALE
1970 PINEGUEST
1971 PINEHURST
1972 PINEISLE
1973 PINELLAS
1974 PINELLI
1975 PINETOP
1976 PINETUCKY
1977 PINEVAL
1978 PINEZ
1979 PINGREE
1980 PINHOOK
1981 PINICON
1982 PINKEL
1983 PINKHAM
1984 PINKSTON
1985 PINNACLES
1986 PINNEBOG
1987 PINNOBIE
1988 PINO
1989 PINOLE
1990 PINON
1991 PINONES
1992 PINTAS
1993 PINTLAR
1994 PINTO
1995 PINTURA
1996 PINTWATER
1997 PIOCHE
1998 PIOPOLIS
1999 PIPELINE
2000 PIPER
2001 PIPESTONE
2002 PIRD
2003 PIRO
2004 PIROUETTE
2005 PIRUM
2006 PISGAH
2007 PISHKUM
2008 PISHKUN
2009 PISMO
2010 PIT
2011 PITCHER
2012 PITCO
2013 PITNEY
2014 PITS
2015 PITS, GRAVEL
2016 PITS, QUARRIES
2017 PITTMAN
2018 PITTSFIELD
2019 PITTSTOWN
2020 PITZER
2021 PIUTE
2022 PIVOT
2023 PIXLEY
2024 PIZENE
2025 PLACAQUODS
2026 PLACEDO
2027 PLACENTIA
2028 PLACERITOS
2029 PLACID
2030 PLACITAS
2031 PLACK
2032 PLAINBO
2033 PLAINFIELD
2034 PLAISTED
2035 PLANK
2036 PLANKINTON
2037 PLANO
2038 PLANTATION
2039 PLASKETT
2040 PLATA
2041 PLATEA
2042 PLATNER
2043 PLATO
2044 PLATORO
2045 PLATTE
2046 PLATTVILLE
2047 PLAYAS
2048 PLAYCO
2049 PLAYER
2050 PLAYMOOR
2051 PLAZA
2052 PLEASANT
2053 PLEASANT GROVE
2054 PLEASANT VALE
2055 PLEASANT VIEW
2056 PLEASANTON
2057 PLEDGER
2058 PLEINE
2059 PLEIOVILLE
2060 PLEITO
2061 PLEVNA
2062 PLITE
2063 PLOME
2064 PLOVER
2065 PLUCK
2066 PLUMAS
2067 PLUMMER
2068 PLUSH
2069 PLUTOS
2070 PLYMOUTH
2071 POARCH
2072 POBER
2073 POCALLA
2074 POCAN
2075 POCASSET
2076 POCATELLO
2077 POCATY
2078 POCKER
2079 POCOLA
2080 POCOMOKE

2081 POCONO
2082 PODEN
2083 PODMOR
2084 PODO
2085 PODUNK
2086 POE
2087 POGAL
2088 POGANEAB
2089 POGUE
2090 POHAKUPU
2091 POIN
2092 POINDEXTER
2093 POINSETT
2094 POINT
2095 POINT ISABEL
2096 POISONCREEK
2097 POJO
2098 POJOAQUE
2099 POKEGEMA
2100 POKEMAN
2101 POKER
2102 POKEY
2103 POLACCA
2104 POLALLIE
2105 POLAR
2106 POLATIS
2107 POLAWANA
2108 POLECREEK
2109 POLELINE
2110 POLEPATCH
2111 POLERUN
2112 POLEY
2113 POLICH
2114 POLKING
2115 POLLARD
2116 POLLUX
2117 POLLY
2118 POLO
2119 POLONIO
2120 POLSON
2121 POLUM
2122 POMAT
2123 POMELLO
2124 POMFRET
2125 POMO
2126 POMONA
2127 POMPANO
2128 POMPTON
2129 POMROY
2130 PONCA
2131 PONCENA
2132 PONCHA
2133 POND
2134 POND CREEK
2135 PONDER
2136 PONIL
2137 PONINA
2138 PONOZZO
2139 PONTO
2140 PONTOTOC
2141 PONZER
2142 POOKU
2143 POOLER
2144 POOLEVILLE
2145 POORCAL
2146 POORMA
2147 POOSE
2148 POOTATUCK
2149 POPASH
2150 POPE
2151 POPHERS
2152 POPLE
2153 POPLIMENTO
2154 POPOSHIA
2155 POPPLETON
2156 POQUITA
2157 POQUONOCK
2158 PORFIRIO
2159 PORRETT
2160 PORT
2161 PORT BYRON
2162 PORTAGE
2163 PORTAGEVILLE
2164 PORTALES
2165 PORTALTO
2166 PORTERFIELD
2167 PORTERS
2168 PORTERVILLE
2169 PORTHILL
2170 PORTIA
2171 PORTINO
2172 PORTLAND
2173 PORTMOUNT
2174 PORTNEUF
2175 PORTOLA
2176 PORTSMOUTH
2177 PORUM
2178 POSANT
2179 POSEY
2180 POSEYVILLE
2181 POSITAS
2182 POSKIN
2183 POSO
2184 POSOS
2185 POST
2186 POTCHUB
2187 POTEET
2188 POTELL
2189 POTH
2190 POTLATCH
2191 POTOMAC
2192 POTOSI
2193 POTRATZ
2194 POTSDAM
2195 POTTER
2196 POTTINGER
2197 POTTS
2198 POTTSBURG
2199 POUDRE
2200 POULSBO
2201 POUNCEY
2202 POVERTY
2203 POVEY
2204 POWDER
2205 POWDERHORN
2206 POWELL
2207 POWER
2208 POWLEY
2209 POWWAHKEE
2210 POY
2211 POYGAN
2212 POYNOR
2213 POZO
2214 POZO BLANCO
2215 PRAG
2216 PRAIRIEVILLE
2217 PRAMISS
2218 PRATHER
2219 PRATLEY
2220 PRATT
2221 PREACHER
2222 PREAKNESS
2223 PREATORSON
2224 PREBISH
2225 PREBLE
2226 PRELO
2227 PREMIER
2228 PRENTISS
2229 PRESA
2230 PRESHER
2231 PRESTO
2232 PRESTON
2233 PREWITT
2234 PREY
2235 PRICE
2236 PRIDA
2237 PRIDHAM
2238 PRIESTLAKE
2239 PRIETA
2240 PRIM
2241 PRIMEAUX
2242 PRIMEN
2243 PRIMGHAR
2244 PRINCETON
2245 PRINEVILLE
2246 PRING
2247 PRINGLE
2248 PRITCHETT
2249 PROCHASKA
2250 PROCTOR
2251 PROGRESSO
2252 PROMD
2253 PROMISE
2254 PROMO
2255 PRONG
2256 PROSPECT
2257 PROSPER
2258 PROSSER
2259 PROTIVIN
2260 PROUT
2261 PROUTY
2262 PROVIDENCE
2263 PROVO
2264 PROVO BAY
2265 PROW
2266 PRUE
2267 PRUITTON
2268 PRUNIE
2269 PRYOR
2270 PSUGA
2271 PTARMIGAN
2272 PUAPUA
2273 PUAULU
2274 PUCHYAN
2275 PUDDLE
2276 PUERCO
2277 PUERTA
2278 PUETT
2279 PUFFER
2280 PUGET
2281 PUGSLEY
2282 PUHI
2283 PUHIMAU
2284 PULA
2285 PULASKI
2286 PULCAN
2287 PULEHU
2288 PULEXAS
2289 PULLMAN
2290 PULPIT
2291 PULS
2292 PULSIPHER
2293 PULTNEY
2294 PUMEL
2295 PUMPER
2296 PUNA
2297 PUNALUU
2298 PUNCHBOWL
2299 PUNDIP
2300 PUNG
2301 PUNGO
2302 PUNOHU
2303 PUNTA
2304 PURCELLA
2305 PURCHES
2306 PURDAM
2307 PURDY
2308 PURETT
2309 PURGATORY
2310 PURNER
2311 PURSLEY
2312 PURVES
2313 PUSHMATAHA
2314 PUSTOI
2315 PUTNAM
2316 PUTNEY
2317 PUTT
2318 PUTTSTER
2319 PUU OO
2320 PUU OPAE
2321 PUU PA
2322 PUUKALA
2323 PUUONE
2324 PUYALLUP
2325 PYBURN
2326 PYLE
2327 PYLON
2328 PYOTE
2329 PYRAMID
2330 PYRMONT
2331 PYWELL
2332 QUAFENO
2333 QUAKER
2334 QUAKERTOWN
2335 QUAM
2336 QUAMON
2337 QUANAH
2338 QUANDER
2339 QUANTICO
2340 QUARLES
2341 QUARTZBURG
2342 QUARTZIPSAMENTS
2343 QUARTZVILLE
2344 QUARZ
2345 QUATAMA
2346 QUAY
2347 QUAZO
2348 QUEALMAN
2349 QUEALY
2350 QUEBRADA
2351 QUEENY
2352 QUEETS
2353 QUEMADO
2354 QUENZER
2355 QUERC
2356 QUERENCIA
2357 QUETICO
2358 QUICKSELL
2359 QUIDEN
2360 QUIENSABE
2361 QUIETUS
2362 QUIGLEY
2363 QUIHI
2364 QUILCENE
2365 QUILLAYUTE
2366 QUILOTOSA
2367 QUILT
2368 QUIMA
2369 QUINCY
2370 QUINCY, MODERATE
2371 QUINLAN
2372 QUINN
2373 QUINNEY
2374 QUINTANA
2375 QUINTO
2376 QUINTON
2377 QUITERIA
2378 QUITMAN
2379 QUIVERA
2380 QUONSET

2381 QUOSATANA
2382 RABBITEX
2383 RABER
2384 RABIDEUX
2385 RABUN
2386 RACE
2387 RACHERT
2388 RACINE
2389 RACKER
2390 RACOMBES
2391 RACOON
2392 RAD
2393 RADDLE
2394 RADER
2395 RADERSBURG
2396 RADFORD
2397 RADLEY
2398 RADNOR
2399 RAFAEL
2400 RAFTON
2401 RAGLAN
2402 RAGNAR
2403 RAGO
2404 RAGPIE
2405 RAGSDALE
2406 RAGTOWN
2407 RAHAL
2408 RAHM
2409 RAHWORTH
2410 RAIL
2411 RAILCITY
2412 RAINBOW
2413 RAINEY
2414 RAINIER
2415 RAINO
2416 RAINS
2417 RAINSBORO
2418 RAIRDENT
2419 RAISIO
2420 RAKE
2421 RAKIED
2422 RALEIGH
2423 RALLOD
2424 RALLS
2425 RALPH
2426 RALPHSTON
2427 RAMADERO
2428 RAMBLA
2429 RAMBOUILLET
2430 RAMELLI
2431 RAMIRES
2432 RAMMEL
2433 RAMO
2434 RAMONA
2435 RAMPART
2436 RAMPARTER
2437 RAMPS
2438 RAMROD
2439 RAMSDELL
2440 RAMSEY
2441 RAMSHORN
2442 RANCE
2443 RANDADO
2444 RANDALL
2445 RANDCORE
2446 RANDMAN
2447 RANDOLPH
2448 RANDS
2449 RANDSBURG
2450 RANGER
2451 RANRUFF
2452 RANSLO
2453 RANSOM
2454 RANSTEIN
2455 RANTOUL
2456 RAPATEE
2457 RAPELJE
2458 RAPH
2459 RAPHO
2460 RAPIDAN
2461 RAPLEE
2462 RAPPAHANNOCK
2463 RAPSON
2464 RARDEN
2465 RARICK
2466 RARITAN
2467 RASBAND
2468 RASILLE
2469 RASSER
2470 RASSET
2471 RATAKE
2472 RATHBUN
2473 RATHDRUM
2474 RATLIFF
2475 RATON
2476 RATSOW
2477 RATTLER
2478 RATTO
2479 RAUB
2480 RAUGHT
2481 RAUVILLE
2482 RAUZI
2483 RAVALLI
2484 RAVEN
2485 RAVENDALE
2486 RAVENELL
2487 RAVENNA
2488 RAVENSWOOD
2489 RAVIA
2490 RAVOLA
2491 RAWAH
2492 RAWE
2493 RAWLES
2494 RAWSON
2495 RAWSONVILLE
2496 RAYBURN
2497 RAYEX
2498 RAYFORD
2499 RAYLAKE
2500 RAYMONDVILLE
2501 RAYNE
2502 RAYNESFORD
2503 RAYNHAM
2504 RAYNOLDSON
2505 RAYPOL
2506 RAZITO
2507 RAZOR
2508 RAZORBA
2509 RAZORT

SOIL05.DAT

1 READING
2 READINGTON
3 READLYN
4 REAGAN
5 REAKOR
6 REAL
7 REAP
8 REARDAN
9 REAVILLE
10 REBA
11 REBEL
12 RECLUSE
13 RED BAY
14 RED BLUFF
15 RED BUTTE
16 RED HOOK
17 RED ROCK
18 RED SPUR
19 REDBANK
20 REDBELL
21 REDBY
22 REDCAN
23 REDCAP
24 REDCHIEF
25 REDCLIFF
26 REDCLOUD
27 REDCO
28 REDCREEK
29 REDDICK
30 REDDING
31 REDEYE
32 REDFEATHER
33 REDFIELD
34 REDIG
35 REDLAKE
36 REDLANDS
37 REDLODGE
38 REDMANSON
39 REDMOND
40 REDNIK
41 REDNUN
42 REDOLA
43 REDONA
44 REDONDO
45 REDPOP
46 REDPORT
47 REDRIDGE
48 REDROB
49 REDROCK
50 REDSPRINGS
51 REDSTOE
52 REDSTONE
53 REDTHAYNE
54 REDTOM
55 REDVALE
56 REDVIEW
57 REDWASH
58 REE
59 REEBOK
60 REED
61 REEDER
62 REEDSBURG
63 REEDSPORT
64 REEDY
65 REEFRIDGE
66 REELFOOT
67 REESE
68 REESVILLE
69 REEVES
70 REFLECTION
71 REFUGE
72 REGAL
73 REGAN
74 REGENT
75 REGGAD
76 REGGEAR
77 REGNAPS
78 REGNIER
79 REHBURG
80 REHFIELD
81 REHM
82 REICHEL
83 REIFF
84 REILLY
85 REINA
86 REINACH
87 REINER
88 REKOP
89 RELAN
90 RELAY
91 RELIANCE
92 RELIZ
93 RELLEY
94 RELSOB
95 RELUCTAN
96 REMBERT
97 REMLAP
98 REMLIK
99 REMMIT
100 REMNOY
101 REMOTE
102 REMSEN
103 REMUNDA
104 REMUS
105 RENBAC
106 RENCALSON
107 RENCOT
108 RENFROW
109 RENICK
110 RENISH
111 RENNER
112 RENNIE
113 RENO
114 RENOHILL
115 RENOL
116 RENOVA
117 RENOX
118 RENSHAW
119 RENSLOW
120 RENSSELAER
121 RENTILL
122 RENTON
123 RENTSAC
124 RENTZEL
125 REPARADA
126 REPP
127 REPPART
128 REPUBLIC
129 RESCUE
130 RESNER
131 RESORT
132 RESOTA
133 RESTING
134 RESTON
135 RET
136 RETRIEVER
137 RETROP
138 REVA
139 REVEL
140 REVENTON
141 REVERE
142 REVIT
143 REWARD
144 REXBURG
145 REXFORD
146 REXMONT
147 REXOR
148 REYAB
149 REYES
150 REYNOSA
151 REYWAT
152 REZAVE
153 RHAME
154 RHEA
155 RHINEBECK
156 RHOADES
157 RHOAME
158 RHOAMETT
159 RHONE
160 RIB
161 RIBERA
162 RIBHILL
163 RICCO
164 RICEBORO
165 RICERT
166 RICETON
167 RICEVILLE
168 RICH
169 RICHARDSON
170 RICHENS
171 RICHEY
172 RICHFIELD
173 RICHFORD
174 RICHLAND
175 RICHMOND
176 RICHTER
177 RICHVALE
178 RICHVIEW
179 RICHVILLE
180 RICHWOOD
181 RICKER
182 RICKMAN
183 RICKMORE
184 RICKREALL
185 RICKS
186 RICOT
187 RICREST
188 RIDD
189 RIDDLES
190 RIDENBAUGH
191 RIDGE
192 RIDGEBURY
193 RIDGECREST
194 RIDGEDALE
195 RIDGELAND
196 RIDGELAWN
197 RIDGEPORT
198 RIDGEVIEW
199 RIDGEVILLE
200 RIDIT
201 RIDOTT
202 RIEDEL
203 RIEDTOWN
204 RIEPE
205 RIESEL
206 RIETBROCK
207 RIFLE
208 RIGA
209 RIGDON
210 RIGGINS
211 RIGGS
212 RIGLEY
213 RIGOLETTE
214 RILEY
215 RILLA
216 RILLINO
217 RILLITO
218 RIMER
219 RIMINI
220 RIMROCK
221 RIMTON
222 RIN
223 RINCON
224 RINDA
225 RINDGE
226 RINEARSON
227 RINEY
228 RING
229 RINGLE
230 RINGLING
231 RINGO
232 RINGWOOD
233 RINKER
234 RIO
235 RIO ARRIBA
236 RIO DIABLO
237 RIO GRANDE
238 RIO LAJAS
239 RIO PIEDRAS
240 RIOCONCHO
241 RIOLINDA
242 RION
243 RIPLEY
244 RIPON
245 RIPPLE
246 RIPPOWAM
247 RIRIE
248 RISBECK
249 RISLEY
250 RISUE
251 RITA
252 RITCHEY
253 RITNER
254 RITO
255 RITTER
256 RITTMAN
257 RITZ
258 RITZCAL
259 RITZVILLE
260 RIVERDALE
261 RIVERHEAD
262 RIVERSIDE
263 RIVERTON
264 RIVERVIEW
265 RIVERWASH
266 RIVIERA
267 RIVRA
268 RIXIE
269 RIZ
270 RIZNO
271 RIZOZO
272 ROANE
273 ROANHIDE
274 ROANOKE
275 ROARING
276 ROB ROY
277 ROBANA
278 ROBBS
279 ROBCO
280 ROBER

281 ROBERTSDALE
282 ROBERTSVILLE
283 ROBIN
284 ROBINETTE
285 ROBINSONVILLE
286 ROBOZO
287 ROBROOST
288 ROBSON
289 ROBY
290 ROCA
291 ROCHE
292 ROCHELLE
293 ROCHER
294 ROCHESTER
295 ROCIO
296 ROCK CREEK
297 ROCK OUTCROP
298 ROCK RIVER
299 ROCKABIN
300 ROCKAWAY
301 ROCKBRIDGE
302 ROCKCASTLE
303 ROCKDALE
304 ROCKERS
305 ROCKERVILLE
306 ROCKFORD
307 ROCKHOUSE
308 ROCKINCHAIR
309 ROCKLIN
310 ROCKLY
311 ROCKOA
312 ROCKTON
313 ROCKWELL
314 ROCKWOOD
315 ROCKY AND STONY
316 ROCKY FORD
317 ROCKYBAR
318 RODAD
319 RODELL
320 RODEO
321 RODESSA
322 RODMAN
323 ROEBUCK
324 ROELLEN
325 ROEMER
326 ROETEX
327 ROFISS
328 ROGAN
329 ROGERSON
330 ROGERT
331 ROGUE
332 ROHNERVILLE
333 ROHRERSVILLE
334 ROIC
335 ROLETTE
336 ROLFE
337 ROLISS
338 ROLLA
339 ROLLINGSTONE
340 ROLOC
341 ROLOFF
342 ROMAYOR
343 ROMBERG
344 ROMBO
345 ROME
346 ROMEO
347 ROMERO
348 ROMGAN
349 ROMIA
350 ROMINE
351 ROMINELL
352 ROMNELL
353 ROMULUS
354 RONAN
355 ROND
356 RONDEAU
357 RONDELL
358 RONNEBY
359 RONSEL
360 RONSON
361 ROONEY
362 ROOSET
363 ROOT
364 ROOTEL
365 ROPER
366 ROQUES
367 ROSALIE
368 ROSAMOND
369 ROSANE
370 ROSANKY
371 ROSARIO
372 ROSCOE
373 ROSCOMMON
374 ROSE CREEK
375 ROSE VALLEY
376 ROSEBERRY
377 ROSEBLOOM
378 ROSEBUD
379 ROSEBURG
380 ROSEDHU
381 ROSEGLEN
382 ROSEHAVEN
383 ROSEHILL
384 ROSELAND
385 ROSELLA
386 ROSELMS
387 ROSENWALL
388 ROSEVILLE
389 ROSEWOOD
390 ROSEWORTH
391 ROSHE SPRINGS
392 ROSHOLT
393 ROSINE
394 ROSITAS
395 ROSLYN
396 ROSMAN
397 ROSNEY
398 ROSS
399 ROSSBURG
400 ROSSFIELD
401 ROSSMOYNE
402 ROSWELL
403 ROSY
404 ROTAMER
405 ROTAN
406 ROTHIEMAY
407 ROTHSAY
408 ROTINOM
409 ROTO
410 ROTTULEE
411 ROUBIDEAU
412 ROUEN
413 ROUGH BROKEN LAN
414 ROUGHCREEK
415 ROUGHMOUNT
416 ROUND BUTTE
417 ROUNDABOUT
418 ROUNDOR
419 ROUNDTOP
420 ROUNDUP
421 ROUNDY
422 ROUSSEAU
423 ROUTON
424 ROUTT
425 ROVAL
426 ROWDEN
427 ROWDY
428 ROWE
429 ROWEL
430 ROWENA
431 ROWLAND
432 ROWLEY
433 ROXAL
434 ROXANA
435 ROXBURY
436 ROXER
437 ROXTON
438 ROY
439 ROYAL
440 ROYAL, BASIN
441 ROYAL, MODERATEL
442 ROYAL, SHALLOW T
443 ROYCE
444 ROYOSA
445 ROYST
446 ROYSTONE
447 ROZA
448 ROZELLVILLE
449 ROZETTA
450 ROZLEE
451 RUARK
452 RUBBLE LAND
453 RUBICON
454 RUBIO
455 RUBSON
456 RUBY
457 RUBYHILL
458 RUCH
459 RUCKER
460 RUCKLES
461 RUCLICK
462 RUDD
463 RUDDLEY
464 RUDEEN
465 RUDYARD
466 RUELLA
467 RUFUS
468 RUGAR
469 RUGLES
470 RUHE
471 RUIDOSO
472 RUKO
473 RULE
474 RUMBLECREEK
475 RUMBO
476 RUMFORD
477 RUMNEY
478 RUMPLE
479 RUMUNG
480 RUNE
481 RUNEBERG
482 RUNGE
483 RUNN
484 RUPLEY
485 RUSCO
486 RUSE
487 RUSH
488 RUSHMORE
489 RUSHTOWN
490 RUSHVILLE
491 RUSO
492 RUSON
493 RUSSELL
494 RUSSIAN
495 RUSSLER
496 RUSTICO
497 RUSTIGATE
498 RUSTON
499 RUSTY
500 RUTAB
501 RUTHERFORD
502 RUTLAND
503 RUTLEGE
504 RYAN
505 RYAN PARK
506 RYARK
507 RYDE
508 RYDER
509 RYDOLPH
510 RYEGATE
511 RYELL
512 RYEPATCH
513 RYER
514 RYKER
515 RYORP
516 RYPOD
517 RYUS
518 SAAR
519 SABANA
520 SABANA SECA
521 SABE
522 SABENYO
523 SABINA
524 SABINE
525 SABLE
526 SAC
527 SACHEEN
528 SACHETT
529 SACKETTS
530 SACO
531 SACRAMENTO
532 SACUL
533 SADDLE
534 SADDLEBACK
535 SADER
536 SADIE
537 SADLER
538 SAFFELL
539 SAG
540 SAGANING
541 SAGASER
542 SAGE
543 SAGECREEK
544 SAGEDALE
545 SAGEHILL
546 SAGEMOOR
547 SAGERS
548 SAGERTON
549 SAGLE
550 SAGO
551 SAGOUSPE
552 SAGUACHE
553 SAILBOAT
554 SAL
555 SALADAR
556 SALADON
557 SALAL
558 SALAMATOF
559 SALANDER
560 SALAS
561 SALCHAKET
562 SALCO
563 SALEM
564 SALERATUS
565 SALERNO
566 SALGA
567 SALIDA
568 SALINAS
569 SALISBURY
570 SALIX
571 SALKUM
572 SALLISAW
573 SALLYANN
574 SALMO
575 SALMON
576 SALONIE
577 SALT CHUCK
578 SALT FLATS
579 SALT LAKE
580 SALTAIR

581 SALTER
582 SALTERY
583 SALTESE
584 SALTINE
585 SALTON
586 SALUDA
587 SALVISA
588 SALZER
589 SAMBA
590 SAMBRITO
591 SAMDAY
592 SAMISH
593 SAMMAMISH
594 SAMOIST
595 SAMOR
596 SAMPSEL
597 SAMPSON
598 SAMSIL
599 SAMSULA
600 SAN ANDREAS
601 SAN ANTON
602 SAN ANTONIO
603 SAN ARCACIO
604 SAN BENITO
605 SAN EMIGDIO
606 SAN GERMAN
607 SAN ISABEL
608 SAN JOAQUIN
609 SAN JON
610 SAN JOSE
611 SAN JUAN
612 SAN LUIS
613 SAN MATEO
614 SAN MIGUEL
615 SAN SABA
616 SAN SEBASTIAN
617 SAN SIMEON
618 SAN TIMOTEO
619 SAN YSIDRO
620 SANCHEZ
621 SANCLEMENTE
622 SANDALL
623 SANDBRANCH
624 SANDCREEK
625 SANDERSON
626 SANDHILL
627 SANDIA
628 SANDOVAL
629 SANDRIDGE
630 SANDUN
631 SANDUSKY
632 SANDWASH
633 SANDWICK
634 SANELI
635 SANFORD
636 SANGER
637 SANGO
638 SANHEDRIN
639 SANIBEL
640 SANILAC
641 SANJE
642 SANLOREN
643 SANPETE
644 SANPITCH
645 SANSARC
646 SANTA
647 SANTA CLARA
648 SANTA FE
649 SANTA ISABEL
650 SANTA LUCIA
651 SANTA MARTA
652 SANTA YNEZ
653 SANTANA
654 SANTANELA
655 SANTAROSA
656 SANTEE
657 SANTIAGO
658 SANTIAM
659 SANTO
660 SANTO TOMAS
661 SANTONI
662 SANWELL
663 SAPELO
664 SAPINERO
665 SAPKIN
666 SAPPHIRE
667 SAPPINGTON
668 SARA
669 SARAGOSA
670 SARAHSVILLE
671 SARALEGUI
672 SARANAC
673 SARAPH
674 SARATON
675 SARAZAN
676 SARBEN
677 SARCILLO
678 SARDINIA
679 SARDIS
680 SARGEANT
681 SARITA
682 SARKAR
683 SARNOSA
684 SARONA
685 SARPY
686 SARTELL
687 SASKA
688 SASPAMCO
689 SASSAFRAS
690 SASSER
691 SATANKA
692 SATANTA
693 SATATTON
694 SATELLITE
695 SATILLA
696 SATIN
697 SATSOP
698 SATT
699 SATTLEY
700 SATTRE
701 SATURN
702 SATUS
703 SAUCEL
704 SAUCIER
705 SAUDE
706 SAUGATUCK
707 SAUGUS
708 SAUK
709 SAULICH
710 SAUM
711 SAUNDERS
712 SAURIN
713 SAUVIE
714 SAUVOLA
715 SAUZ
716 SAVAGE
717 SAVAGETON
718 SAVANNAH
719 SAVENAC
720 SAVO
721 SAVOIA
722 SAVONA
723 SAVONIA
724 SAWABE
725 SAWATCH
726 SAWBUCK
727 SAWCREEK
728 SAWDUST
729 SAWLIT
730 SAWMILL
731 SAWTELL
732 SAWTELPEAK
733 SAWYER
734 SAXBY
735 SAXON
736 SAY
737 SAYBROOK
738 SAYDAB
739 SAYERS
740 SAYLES
741 SAYLESVILLE
742 SAYNER
743 SAYPO
744 SAZI
745 SCALA
746 SCALLEY
747 SCAMMAN
748 SCANDARD
749 SCANTIC
750 SCANTIC VARIANT
751 SCAPONIA
752 SCAR
753 SCARBORO
754 SCATLAKE
755 SCAVE
756 SCHAFFENAKER
757 SCHALLER
758 SCHAMBER
759 SCHAMP
760 SCHAPVILLE
761 SCHAWANA
762 SCHENCO
763 SCHERRARD
764 SCHLEY
765 SCHMUTZ
766 SCHNEBLY
767 SCHNEIDER
768 SCHNIPPER
769 SCHNOORSON
770 SCHNORBUSH
771 SCHODSON
772 SCHOFIELD
773 SCHOHARIE
774 SCHOLLE
775 SCHOODIC
776 SCHOOLCRAFT
777 SCHOOLEY
778 SCHOOLHOUSE
779 SCHOONER
780 SCHRADER
781 SCHRAP
782 SCHRIER
783 SCHROCK
784 SCHROON
785 SCHUELKE
786 SCHULINE
787 SCHUMACHER
788 SCHUSTER
789 SCHUYLER
790 SCHWANA
791 SCIO
792 SCIOTOVILLE
793 SCISM
794 SCITICO
795 SCITUATE
796 SCOAP
797 SCOBEY
798 SCOGGIN
799 SCOON
800 SCOOTENEY
801 SCORIA LAND
802 SCORUP
803 SCOTCO
804 SCOTIA
805 SCOTT
806 SCOTT LAKE
807 SCOTTIES
808 SCOUT
809 SCRABBLERS
810 SCRANTON
811 SCRAVO
812 SCRIBA
813 SCRIVER
814 SCROGGIN
815 SCULLIN
816 SCUPPERNONG
817 SEABROOK
818 SEAFIELD
819 SEAFORTH
820 SEAGATE
821 SEAGOVILLE
822 SEALY
823 SEAMAN
824 SEAQUEST
825 SEAR
826 SEARING
827 SEARLA
828 SEARLES
829 SEARSPORT
830 SEARSVILLE
831 SEASTRAND
832 SEATON
833 SEATTLE
834 SEAVERSON
835 SEAWILLOW
836 SEBAGO
837 SEBASTIAN
838 SEBASTOPOL
839 SEBEWA
840 SEBREE
841 SEBRING
842 SEBUD
843 SECATA
844 SECCA
845 SECESH
846 SECRET CREEK
847 SEDALE
848 SEDGEFIELD
849 SEDGEWAY
850 SEDILLO
851 SEDMAR
852 SEDROWOOLLEY
853 SEDWELL
854 SEEDSKADEE
855 SEELEZ
856 SEELOVERS
857 SEELYEVILLE
858 SEEPRID
859 SEES
860 SEEWEE
861 SEFFNER
862 SEGIDAL
863 SEGNO
864 SEGUIN
865 SEGURA
866 SEHOME
867 SEHORN
868 SEIS
869 SEITZ
870 SEJITA
871 SEKIL
872 SEKIU
873 SELAH
874 SELDEN
875 SELEVIN
876 SELFRIDGE
877 SELIA
878 SELIGMAN
879 SELKIRK
880 SELLE

881 SELLERS
882 SELMA
883 SELMAC
884 SELON
885 SELTI
886 SELWAY
887 SEMIAHMOO
888 SEMINOLE
889 SEN
890 SENCHERT
891 SENECAVILLE
892 SENSABAUGH
893 SEQUATCHIE
894 SEQUIM
895 SEQUOIA
896 SERDEN
897 SERENE
898 SEROCO
899 SERPA
900 SERPEN
901 SERPENTANO
902 SERPOD
903 SERRANO
904 SERVILLETA
905 SESAME
906 SESPE
907 SESSIONS
908 SESSUM
909 SET
910 SETH
911 SETTERS
912 SETTLEMENT
913 SETTLEMEYER
914 SEVAL
915 SEVENMILE
916 SEVERN
917 SEVILLE
918 SEVY
919 SEWANEE
920 SEWARD
921 SEXTON
922 SEYMOUR
923 SHAAK
924 SHABLISS
925 SHACK
926 SHADELAND
927 SHADELEAF
928 SHADOW
929 SHADYGROVE
930 SHAFFTON
931 SHAGEL
932 SHAGNASTY
933 SHAKAMAK
934 SHAKER
935 SHAKESPEARE
936 SHAKOPEE
937 SHALAKE
938 SHALAKO
939 SHALBA
940 SHALCAR
941 SHALCR
942 SHALET
943 SHALONA
944 SHAM
945 SHAMBO
946 SHAMEL
947 SHAMOCK
948 SHANAHAN
949 SHANDEP
950 SHANE
951 SHANGHAI
952 SHANKLER
953 SHANO
954 SHANTA
955 SHARATIN
956 SHARKEY
957 SHARLAND
958 SHARON
959 SHARPS
960 SHARPSBURG
961 SHARROTT
962 SHARVANA
963 SHASER
964 SHASKIT
965 SHASTA
966 SHATRUCE
967 SHATTA
968 SHATTUCK
969 SHAVANO
970 SHAVASH
971 SHAVER
972 SHAWA
973 SHAWANO
974 SHAWMUT
975 SHAY
976 SHAYLA
977 SHEAR
978 SHEAVILLE
979 SHEBANG
980 SHEBEON
981 SHEDADO
982 SHEDD
983 SHEDHORN
984 SHEEGE
985 SHEEK
986 SHEEP CREEK
987 SHEEPCAN
988 SHEEPHEAD
989 SHEEPROCK
990 SHEEPSCOT
991 SHEETIRON
992 SHEFFIELD
993 SHELBURNE
994 SHELBY
995 SHELBYVILLE
996 SHELD
997 SHELL
998 SHELLABARGER
999 SHELLBLUFF
1000 SHELLDRAKE
1001 SHELLROCK
1002 SHELMADINE
1003 SHELOCTA
1004 SHELOCTA VARIANT
1005 SHELTON
1006 SHENA
1007 SHENANDOAH
1008 SHENKS
1009 SHENVAL
1010 SHEP
1011 SHEPAN
1012 SHEPPARD
1013 SHERANDO
1014 SHERAR
1015 SHERBURNE
1016 SHERIDAN
1017 SHERLESS
1018 SHERLOCK
1019 SHERM
1020 SHERMORE
1021 SHERRY
1022 SHERRYL
1023 SHERWOOD
1024 SHEVLIN
1025 SHIDLER
1026 SHIELDS
1027 SHIFFER
1028 SHILOH
1029 SHIMA
1030 SHIMMON
1031 SHINAKU
1032 SHINBARA
1033 SHINDLER
1034 SHINER
1035 SHINGLE
1036 SHINGLETOWN
1037 SHINKEE
1038 SHINROCK
1039 SHIOCTON
1040 SHIPLEY
1041 SHIPROCK
1042 SHIPS
1043 SHIPSHE
1044 SHIRK
1045 SHIRLEY
1046 SHIVELY
1047 SHOALS
1048 SHOAT
1049 SHOBA
1050 SHOEPEG
1051 SHOESTRING
1052 SHOKEN
1053 SHONKIN
1054 SHOOFLIN
1055 SHOOFLY
1056 SHOOK
1057 SHOOKER
1058 SHOREEK
1059 SHOREWOOD
1060 SHORIM
1061 SHORT CREEK
1062 SHORTCUT
1063 SHORTHORN
1064 SHORTYORK
1065 SHOSHONE
1066 SHOTGUN
1067 SHOTWELL
1068 SHOUNS
1069 SHOUSE
1070 SHOWALTER
1071 SHOWLOW
1072 SHREE
1073 SHREWDER
1074 SHREWSBURY
1075 SHRINE
1076 SHROE
1077 SHROUTS
1078 SHUBUTA
1079 SHUE
1080 SHUKASH
1081 SHUKSAN
1082 SHULE
1083 SHULLSBURG
1084 SHUMLA
1085 SHUMWAY
1086 SHUPERT
1087 SHURLEY
1088 SHUSTER
1089 SI
1090 SIBELIA
1091 SIBLEY
1092 SIBLEYVILLE
1093 SICKLES
1094 SICKLESTEETS
1095 SIDDOWAY
1096 SIDELL
1097 SIDLAKE
1098 SIDON
1099 SIEBEN
1100 SIEBERT
1101 SIECHE
1102 SIELO
1103 SIEROCLIFF
1104 SIERRA
1105 SIERRAVILLE
1106 SIESTA
1107 SIFTON
1108 SIGNAL
1109 SIGURD
1110 SIKESTON
1111 SILAS
1112 SILAWA
1113 SILENT
1114 SILER
1115 SILERTON
1116 SILHOUETTE
1117 SILI
1118 SILSTID
1119 SILVA
1120 SILVER
1121 SILVER CREEK
1122 SILVERADO
1123 SILVERBELL
1124 SILVERBOW
1125 SILVERCHIEF
1126 SILVERCLIFF
1127 SILVERDALE
1128 SILVERN
1129 SILVERTON
1130 SILVIES
1131 SIMAS
1132 SIMCOE
1133 SIMEON
1134 SIMEROI
1135 SIMMONT
1136 SIMON
1137 SIMONA
1138 SIMONTON
1139 SIMPARK
1140 SIMPATICO
1141 SIMPSON
1142 SIMS
1143 SINAI
1144 SINAMOX
1145 SINCLAIR
1146 SINGATSE
1147 SINGERTON
1148 SINGLETREE
1149 SINGSAAS
1150 SINKSON
1151 SINLOC
1152 SINNICE
1153 SINNIGAM
1154 SINTON
1155 SINUK
1156 SION
1157 SIOUX
1158 SIOUXON
1159 SIPPLE
1160 SIPSEY
1161 SIRDRAK
1162 SIRI
1163 SIROCO
1164 SIRRETTA
1165 SISKIYOU
1166 SISSETON
1167 SISSON
1168 SISTERS
1169 SITES
1170 SIWELL
1171 SIXBEACON
1172 SIXMILE
1173 SIZER
1174 SKAGGS
1175 SKAGIT
1176 SKAHA
1177 SKALAN
1178 SKAMANIA
1179 SKAMO
1180 SKANEE

1181 SKANID
1182 SKATE
1183 SKEDADDLE
1184 SKELLOCK
1185 SKELON
1186 SKERRY
1187 SKIDMORE
1188 SKIPANON
1189 SKIPOPA
1190 SKIYOU
1191 SKOKOMISH
1192 SKOLY
1193 SKOOKUM
1194 SKOS
1195 SKOWHEGAN
1196 SKULL CREEK
1197 SKUMPAH
1198 SKUTUM
1199 SKYBERG
1200 SKYHAVEN
1201 SKYHIGH
1202 SKYKOMISH
1203 SKYLICK
1204 SKYLINE
1205 SKYMOR
1206 SKYVILLAGE
1207 SKYWAY
1208 SLABTOWN
1209 SLAGLE
1210 SLAUGHTER
1211 SLAUGHTERVILLE
1212 SLAVEN
1213 SLAW
1214 SLAYTON
1215 SLEEPER
1216 SLEETH
1217 SLICKENS
1218 SLICKROCK
1219 SLICKSPOTS
1220 SLIDELL
1221 SLIGHTS
1222 SLIGTING
1223 SLIKOK
1224 SLIMBUTTE
1225 SLINGER
1226 SLIPMAN
1227 SLOAN
1228 SLOCUM
1229 SLUICE
1230 SLY
1231 SMACKOUT
1232 SMALL
1233 SMALLCONE
1234 SMARTS
1235 SMAUG
1236 SMEDLEY
1237 SMELTER
1238 SMILEY
1239 SMILEYVILLE
1240 SMILO
1241 SMITHBORO
1242 SMITHDALE
1243 SMITHNECK
1244 SMITHNECK SERIES
1245 SMITHTON
1246 SMITHVILLE
1247 SMITHWICK
1248 SMOCREEK
1249 SMOKEY
1250 SMOLAN
1251 SMYRNA
1252 SNAG
1253 SNAHOPISH
1254 SNAKE
1255 SNAKE HOLLOW
1256 SNAKELUM
1257 SNAPP
1258 SNEAD
1259 SNEFFELS
1260 SNELL
1261 SNELLING
1262 SNIDER
1263 SNOHOMISH
1264 SNOMO
1265 SNOOK
1266 SNOPOC
1267 SNOQUALMIE
1268 SNOW
1269 SNOWLIN
1270 SNOWMORE
1271 SNOWSHOE
1272 SNOWSLIDE
1273 SNOWSTORM
1274 SNOWVILLE
1275 SNUFFUL
1276 SOAKPAK
1277 SOAPCREEK
1278 SOBEGA
1279 SOBOBA
1280 SOBOL
1281 SOBRANTE
1282 SOCORRO
1283 SODA
1284 SODA LAKE
1285 SODABAY
1286 SODERVILLE
1287 SODHOUSE
1288 SODUS
1289 SOELBERG
1290 SOEN
1291 SOFIA
1292 SOFTSCRABBLE
1293 SOGI
1294 SOGN
1295 SOGO
1296 SOGZIE
1297 SOLAK
1298 SOLANO
1299 SOLDATNA
1300 SOLDIER
1301 SOLDUC
1302 SOLEDAD
1303 SOLIER
1304 SOLIS
1305 SOLLEKS
1306 SOLLER
1307 SOLOMON
1308 SOLONA
1309 SOMBORDORO
1310 SOMBRERO
1311 SOMERS
1312 SOMERVELL
1313 SOMSEN
1314 SONAHNPIL
1315 SONOCAN
1316 SONOITA
1317 SONOMA
1318 SONORA
1319 SONTAG
1320 SOOLAKE
1321 SOONAHBE
1322 SOOSAP
1323 SOPER
1324 SOQUEL
1325 SORENSEN
1326 SORF
1327 SORRENTO
1328 SORTER
1329 SORUM
1330 SOSA
1331 SOSTIEN
1332 SOTIM
1333 SOUGHE
1334 SOULAJULE
1335 SOUTHACE
1336 SOUTHAM
1337 SOUTHFORK
1338 SOUTHGATE
1339 SOUTHPORT
1340 SOUTHRIDGE
1341 SOUTHWICK
1342 SOWCAN
1343 SPAA
1344 SPACE CITY
1345 SPADE
1346 SPADRA
1347 SPAGER
1348 SPALDING
1349 SPANA
1350 SPANAWAY
1351 SPANEL
1352 SPANG
1353 SPANGLER
1354 SPARANK
1355 SPARHAM
1356 SPARKHULE
1357 SPARMO
1358 SPARR
1359 SPARTA
1360 SPASPREY
1361 SPEAKER
1362 SPEAKS
1363 SPEARFISH
1364 SPEARHEAD
1365 SPEARMAN
1366 SPEARVILLE
1367 SPECK
1368 SPEELYAI
1369 SPEER
1370 SPEIGLE
1371 SPENARD
1372 SPENCER
1373 SPENLO
1374 SPENS
1375 SPERRY
1376 SPEXARTH
1377 SPHINX
1378 SPICER
1379 SPICERTON
1380 SPICEWOOD
1381 SPIKE
1382 SPILLCO
1383 SPILLVILLE
1384 SPINEKOP
1385 SPINKS
1386 SPINLIN
1387 SPIRES
1388 SPIRIT
1389 SPIRO
1390 SPIVEY
1391 SPLENDORA
1392 SPLITRO
1393 SPLITTOP
1394 SPOFFORD
1395 SPOFMORE
1396 SPOKANE
1397 SPOKEL
1398 SPONSELLER
1399 SPOOL
1400 SPOONER
1401 SPOTSYLVANIA
1402 SPOTTSWOOD
1403 SPRAY
1404 SPRECKELS
1405 SPRING
1406 SPRINGDALE
1407 SPRINGER
1408 SPRINGERVILLE
1409 SPRINGFIELD
1410 SPRINGMEYER
1411 SPRINGSTEEN
1412 SPRINGWATER
1413 SPROUL
1414 SPRUCEDALE
1415 SPUD
1416 SPUKWUSH
1417 SPUR
1418 SPURGER
1419 SPURLOCA
1420 SPURLOCK
1421 SQUALICUM
1422 SQUALLY
1423 SQUAW
1424 SQUAWCREEK
1425 SQUAWROCK
1426 SQUAWTIP
1427 SQUIRES
1428 SRIADA
1429 ST. ALBANS
1430 ST. ANTHONY
1431 ST. AUGUSTINE
1432 ST. CHARLES
1433 ST. CLAIR
1434 ST. ELMO
1435 ST. GEORGE
1436 ST. HELENS
1437 ST. IGNACE
1438 ST. JOHNS
1439 ST. LUCIE
1440 ST. MARTIN
1441 ST. MARYS
1442 ST. ONGE
1443 ST. PAUL
1444 ST. THOMAS
1445 STABLER
1446 STADY
1447 STAFFORD
1448 STAGECOACH
1449 STAHL
1450 STAKE
1451 STALEY
1452 STALLINGS
1453 STAMBAUGH
1454 STAMFORD
1455 STAMP
1456 STAMPEDE
1457 STAN
1458 STANDLEY
1459 STANDUP
1460 STANEY
1461 STANFIELD
1462 STANISLAUS
1463 STAPALOOP
1464 STAPLETON
1465 STAPP
1466 STARBUCK
1467 STARGO
1468 STARHOPE
1469 STARICHKOF
1470 STARKEY
1471 STARKS
1472 STARLEY
1473 STARMAN
1474 STARR
1475 STASER
1476 STATE
1477 STATELINE
1478 STATLER
1479 STATZ
1480 STAVE

1481 STAVELY
1482 STAYTON
1483 STEARNS
1484 STECUM
1485 STEED
1486 STEEDMAN
1487 STEEKEE
1488 STEELE
1489 STEENS
1490 STEEPCAN
1491 STEESE
1492 STEEVER
1493 STEFF
1494 STEGALL
1495 STEIGER
1496 STEINAUER
1497 STEINBECK
1498 STEINSBURG
1499 STEIWER
1500 STELLAR
1501 STEMBER
1502 STEMILT
1503 STEMLEY
1504 STEMPLE
1505 STENDAL
1506 STEPHEN
1507 STEPHENVILLE
1508 STEPROCK
1509 STEPSTONE
1510 STEPTOE
1511 STERLING
1512 STERLINGTON
1513 STERRETT
1514 STETSON
1515 STETTER
1516 STEUBEN
1517 STEUBER
1518 STEVENS
1519 STEVENSON
1520 STEWART
1521 STEWVAL
1522 STICKNEY
1523 STIDHAM
1524 STIEN
1525 STIGLER
1526 STILGAR
1527 STILL
1528 STILLMAN
1529 STILLWATER
1530 STILSKIN
1531 STILSON
1532 STIMCA
1533 STIMSON
1534 STINGAL
1535 STINGDORN
1536 STIPE
1537 STIRK
1538 STIRRUP
1539 STIRUM
1540 STISSING
1541 STIVERSVILLE
1542 STOCKADE
1543 STOCKBRIDGE
1544 STOCKEL
1545 STOCKLAND
1546 STOCKPEN
1547 STOCKTON
1548 STODA
1549 STODICK
1550 STOHLMAN
1551 STOKES
1552 STOKLY
1553 STOMAR
1554 STONEBERGER
1555 STONEBURG
1556 STONEHAM
1557 STONEHEAD
1558 STONELICK
1559 STONER
1560 STONEVILLE
1561 STONEWALL
1562 STONEWELL
1563 STONO
1564 STONY AND ROCKY
1565 STONYFORD
1566 STOOKMOOR
1567 STORDEN
1568 STORLA
1569 STORMITT
1570 STOTT
1571 STOUGH
1572 STOUT
1573 STOVHO
1574 STOWE
1575 STOWELL
1576 STOY
1577 STRABER
1578 STRAHAN
1579 STRAIGHT
1580 STRANDQUIST
1581 STRAT
1582 STRATFORD
1583 STRATTON
1584 STRAW
1585 STRAWN
1586 STREATOR
1587 STREVELL
1588 STRICKER
1589 STRICKLAND
1590 STRINGTOWN
1591 STROLE
1592 STROM
1593 STROMAL
1594 STRONGHURST
1595 STROUPE
1596 STROZI
1597 STRYCH
1598 STRYKER
1599 STUBBS
1600 STUCKY
1601 STUKEL
1602 STUMBLE
1603 STUMPP
1604 STUMPTOWN
1605 STUNNER
1606 STUNTZ
1607 STURGEON
1608 STURKIE
1609 STUTTGART
1610 STUTZMAN
1611 STUTZVILLE
1612 STYX
1613 SUBACO
1614 SUBLETTE
1615 SUBLIGNA
1616 SUCARNOOCHEE
1617 SUCCESS
1618 SUCHES
1619 SUCOLINE
1620 SUDBURY
1621 SUDDUTH
1622 SUDLEY
1623 SUDWORTH
1624 SUEPERT
1625 SUEY
1626 SUFFIELD
1627 SUFFOLK
1628 SUGARBOWL
1629 SUGARDEE
1630 SUGARLOAF
1631 SUGLO
1632 SUISUN
1633 SULA
1634 SULFAQUEPTS
1635 SULLIVAN
1636 SULLY
1637 SULOAF
1638 SULPHURA
1639 SULSAVAR
1640 SULTAN
1641 SUMAN
1642 SUMAS
1643 SUMATRA
1644 SUMINE
1645 SUMMERFIELD
1646 SUMMERS
1647 SUMMERTON
1648 SUMMERVILLE
1649 SUMMIT
1650 SUMMITVILLE
1651 SUMPF
1652 SUMTER
1653 SUMTERVILLE
1654 SUMYA
1655 SUN
1656 SUNAPEE
1657 SUNBURG
1658 SUNBURST
1659 SUNBURY
1660 SUNCITY
1661 SUNCOOK
1662 SUNDANCE
1663 SUNDAY
1664 SUNDELL
1665 SUNDOWN
1666 SUNEV
1667 SUNFIELD
1668 SUNLIGHT
1669 SUNNYHAY
1670 SUNNYSIDE
1671 SUNNYVALE
1672 SUNRAY
1673 SUNSET
1674 SUNSHINE
1675 SUNSWEET
1676 SUNUP
1677 SUOMI
1678 SUP
1679 SUPAN
1680 SUPERIOR
1681 SUPERSTITION
1682 SUPERVISOR
1683 SUPPLEE
1684 SUR
1685 SURFSIDE
1686 SURGEM
1687 SURGH
1688 SURPRISE
1689 SURRENCY
1690 SURRETT
1691 SURVEYORS
1692 SURVYA
1693 SUSANNA
1694 SUSIE CREEK
1695 SUSITNA
1696 SUSQUEHANNA
1697 SUTA
1698 SUTCLIFF
1699 SUTHER
1700 SUTHERLIN
1701 SUTKIN
1702 SUTLEY
1703 SUTPHEN
1704 SUTRO
1705 SUTTLER
1706 SUTTON
1707 SUVER
1708 SVEA
1709 SVENSEN
1710 SVERDRUP
1711 SWAGER
1712 SWAKANE
1713 SWALER
1714 SWAMP
1715 SWAMPYDRAW
1716 SWANBOY
1717 SWANDAD
1718 SWANLAKE
1719 SWANNER
1720 SWANSEA
1721 SWANSON
1722 SWANTON
1723 SWANTOWN
1724 SWANVILLE
1725 SWANWICK
1726 SWAPPS
1727 SWARTSWOOD
1728 SWARTZ
1729 SWASEY
1730 SWASTIKA
1731 SWAUK
1732 SWEATMAN
1733 SWEDE
1734 SWEEN
1735 SWEENEY
1736 SWEET
1737 SWEETAPPLE
1738 SWEETGRASS
1739 SWEETWATER
1740 SWEITBERG
1741 SWEITING
1742 SWEM
1743 SWENODA
1744 SWIFT
1745 SWIFT CREEK
1746 SWIFTON
1747 SWIMLEY
1748 SWIMS
1749 SWINGLER
1750 SWINK
1751 SWINOMISH
1752 SWINT
1753 SWISBOB
1754 SWISSHELM
1755 SWITCHBACK
1756 SWITZERLAND
1757 SWOPE
1758 SWORMVILLE
1759 SWYGERT
1760 SYBLON
1761 SYCAMORE
1762 SYCAN
1763 SYCLE
1764 SYENITE
1765 SYLACAUGA
1766 SYLCO
1767 SYLVAN
1768 SYLVESTER
1769 SYLVIA
1770 SYMCO
1771 SYMERTON
1772 SYNAREP
1773 SYRACUSE
1774 SYRENE
1775 SYRETT
1776 TABECHEDING
1777 TABERNASH
1778 TABLE MOUNTAIN
1779 TABLER
1780 TABOR

1781 TACAN
1782 TACOMA
1783 TACONIC
1784 TACOOSH
1785 TADLOCK
1786 TAFOYA
1787 TAFT
1788 TAFUNA
1789 TAGGART
1790 TAGLAKE
1791 TAHKENITCH
1792 TAHOMA
1793 TAHOULA
1794 TAHQUATS
1795 TAINTOR
1796 TAJILA
1797 TAJO
1798 TAKEUCHI
1799 TAKILMA
1800 TAKOTNA
1801 TALAG
1802 TALANTE
1803 TALAPUS
1804 TALBOTT
1805 TALCO
1806 TALCOT
1807 TALIHINA
1808 TALKEETNA
1809 TALLA
1810 TALLAC
1811 TALLADEGA
1812 TALLAPOOSA
1813 TALLEYVILLE
1814 TALLS
1815 TALLULA
1816 TALLY
1817 TALMAGE
1818 TALMO
1819 TALMOON
1820 TALOKA
1821 TALPA
1822 TALQUIN
1823 TAMA
1824 TAMAHA
1825 TAMALCO
1826 TAMALPAIS
1827 TAMANEEN
1828 TAMBA
1829 TAMELY
1830 TAMFLAT
1831 TAMFORD
1832 TAMMANY CREEK
1833 TAMMING
1834 TAMP
1835 TAMPICO
1836 TANAK
1837 TANAMA
1838 TANANA
1839 TANBARK
1840 TANDY
1841 TANEUM
1842 TANEY
1843 TANGAIR
1844 TANGLE
1845 TANNA
1846 TANNAHILL
1847 TANNER
1848 TANOB
1849 TANSEM
1850 TANTALUS
1851 TANTILE
1852 TANWAX
1853 TANYARD
1854 TAPCO
1855 TAPIA
1856 TAPICITOES
1857 TAPPAN
1858 TARA
1859 TARBORO
1860 TARGHEE
1861 TARKINGTON
1862 TARKIO
1863 TARKLIN
1864 TARLOC
1865 TARNAV
1866 TARPLEY
1867 TARR
1868 TARRANT
1869 TARRETE
1870 TARRYALL
1871 TARRYTOWN
1872 TASAYA
1873 TASCOSA
1874 TASSEL
1875 TASSELMAN
1876 TASSO
1877 TATE
1878 TATIYEE
1879 TATLUM
1880 TATOUCHE
1881 TATTON
1882 TATUM
1883 TAUNTON
1884 TAVARES
1885 TAWAH
1886 TAWAS
1887 TAWCAW
1888 TAYLOR
1889 TAYLOR CREEK
1890 TAYLORSFLAT
1891 TAYLORSVILLE
1892 TAZLINA
1893 TEAGULF
1894 TEAKEAN
1895 TEALSON
1896 TEALWHIT
1897 TEANAWAY
1898 TEAPO
1899 TEASDALE
1900 TEBO
1901 TECHICK
1902 TECO
1903 TECOLOTE
1904 TECOPA
1905 TEDROW
1906 TEEL
1907 TEELER
1908 TEEMAT
1909 TEETERS
1910 TEEWINOT
1911 TEFTON
1912 TEGURO
1913 TEHACHAPI
1914 TEHAMA
1915 TEHRAN
1916 TEIGEN
1917 TEJA
1918 TEJABE
1919 TEJANA
1920 TEKENINK
1921 TEKISON
1922 TEKLANIKA
1923 TEKOA
1924 TELA
1925 TELCHER
1926 TELEFONO
1927 TELEPHONE
1928 TELESCOPE
1929 TELFER
1930 TELFERNER
1931 TELL
1932 TELLER
1933 TELLICO
1934 TELLMAN
1935 TELLURA
1936 TELOS
1937 TELSTAD
1938 TEMAN
1939 TEMBLOR
1940 TEMESCAL
1941 TEMO
1942 TEMPLE
1943 TEMPLETON
1944 TEMVIK
1945 TENABO
1946 TENAHA
1947 TENAS
1948 TENCEE
1949 TENDOY
1950 TENERIFFE
1951 TENEX
1952 TENINO
1953 TENMILE
1954 TENNO
1955 TENORIO
1956 TENOT
1957 TENPIN
1958 TENRAG
1959 TENSAS
1960 TENSED
1961 TENSLEEP
1962 TENSNOIR
1963 TENVORRD
1964 TEOCULLI
1965 TEPETE
1966 TEQUESTA
1967 TERADA
1968 TERBIES
1969 TERENCE
1970 TERESA
1971 TERINO
1972 TERLCO
1973 TERLINGUA
1974 TERMINAL
1975 TERMO
1976 TEROMOTE
1977 TEROUGE
1978 TERRA CEIA
1979 TERRAD
1980 TERRETON
1981 TERRIC BOROFIBRI
1982 TERRIC BOROSAPRI
1983 TERRIC CRYOFIBRI
1984 TERRIC MEDASAPRI
1985 TERRIL
1986 TERRO
1987 TERRY
1988 TERWILLIGER
1989 TESAJO
1990 TESSFIVE
1991 TETHRICK
1992 TETON
1993 TETONIA
1994 TETONKA
1995 TETONVIEW
1996 TETONVILLE
1997 TETOTUM
1998 TEVIS
1999 TEW
2000 TEWA
2001 TEX
2002 TEXARK
2003 TEXLINE
2004 TEXROY
2005 TEZUMA
2006 THACKER
2007 THACKERY
2008 THADER
2009 THAGE
2010 THATCHER
2011 THATUNA
2012 THAYNE
2013 THEBES
2014 THEBO
2015 THEDALUND
2016 THEEDLE
2017 THENAS
2018 THEODOR
2019 THEON
2020 THERESA
2021 THERIOT
2022 THERMOPOLIS
2023 THESS
2024 THETFORD
2025 THETIS
2026 THIEFRIVER
2027 THIEL
2028 THIESSEN
2029 THIKE
2030 THIOKOL
2031 THIRST
2032 THISTLEDEW
2033 THOENY
2034 THOMAS
2035 THOMS
2036 THORNBURGH
2037 THORNDALE
2038 THORNDIKE
2039 THORNOCK
2040 THORNTON
2041 THOROUGHFARE
2042 THORP
2043 THOUT
2044 THOW
2045 THOWSON
2046 THRASH
2047 THREADGILL
2048 THREECHOP
2049 THREEK
2050 THREEMILE
2051 THROCK
2052 THULEPAH
2053 THUMBERLAND
2054 THUNDERBIRD
2055 THUNDERDRA
2056 THURBER
2057 THURLONI
2058 THURLOW
2059 THURMAN
2060 THURMONT
2061 THWOOP
2062 TIAGOS
2063 TIAK
2064 TIBAN
2065 TIBBITTS
2066 TIBS
2067 TIBURONES
2068 TICA
2069 TICE
2070 TICELL
2071 TICHNOR
2072 TICINO
2073 TICKAPOO
2074 TICKASON
2075 TIDINGS
2076 TIDWELL
2077 TIERRA
2078 TIETON
2079 TIFFANY
2080 TIFTON

2081 TIGER CREEK
2082 TIGERON
2083 TIGIT
2084 TIGIWON
2085 TIGON
2086 TIGUA
2087 TIJERAS
2088 TIKI
2089 TILFER
2090 TILFORD
2091 TILLEDA
2092 TILLICUM
2093 TILLMAN
2094 TILLOU
2095 TILMA
2096 TILSIT
2097 TILTON
2098 TIMBALIER
2099 TIMBERG
2100 TIMBERHEAD
2101 TIMBERVILLE
2102 TIMBLIN
2103 TIMENTWA
2104 TIMHILL
2105 TIMKEN
2106 TIMMERMAN
2107 TIMOR
2108 TIMPAHUTE
2109 TIMPANOGOS
2110 TIMPER
2111 TIMULA
2112 TINA
2113 TINAJA
2114 TINAMOU
2115 TINDAHAY
2116 TINE
2117 TINEMAN
2118 TINGEY
2119 TINKER
2120 TINN
2121 TINNIN
2122 TINSLEY
2123 TINTON
2124 TINYTOWN
2125 TIOCANO
2126 TIOGA
2127 TIPPAH
2128 TIPPECANOE
2129 TIPPER
2130 TIPPERARY
2131 TIPPIPAH
2132 TIPPO
2133 TIPTON
2134 TIPTONVILLE
2135 TIPTOP
2136 TIRO
2137 TISBURY
2138 TISCH
2139 TISDALE
2140 TISHAR
2141 TISONIA
2142 TISWORTH
2143 TITUS
2144 TITUSVILLE
2145 TIVOLI
2146 TIVY
2147 TOA
2148 TOADLAKE
2149 TOBICO
2150 TOBIN
2151 TOBISH
2152 TOBLER
2153 TOBOSA
2154 TOBY
2155 TOCAL
2156 TOCALOMA
2157 TOCAN
2158 TOCCOA
2159 TOCK
2160 TOCOI
2161 TODDLER
2162 TODDSTAV
2163 TODDVILLE
2164 TODOS
2165 TOEHEAD
2166 TOEJA
2167 TOEM
2168 TOGNONI
2169 TOGO
2170 TOGUS
2171 TOHONA
2172 TOIMI
2173 TOINE
2174 TOISNOT
2175 TOIYABE
2176 TOKLAT
2177 TOKOPER
2178 TOKUL
2179 TOLANY
2180 TOLBY
2181 TOLEDO
2182 TOLEX
2183 TOLICHA
2184 TOLKE
2185 TOLL
2186 TOLLGATE
2187 TOLLHOUSE
2188 TOLMAN
2189 TOLNA
2190 TOLO
2191 TOLONIER
2192 TOLOVANA
2193 TOLSONA
2194 TOLSTOI
2195 TOLTEC
2196 TOLUCA
2197 TOLVAR
2198 TOMAH
2199 TOMAHAWK
2200 TOMALES
2201 TOMASAKI
2202 TOMAST
2203 TOMBAR
2204 TOME
2205 TOMEL
2206 TOMERA
2207 TOMICHI
2208 TOMOKA
2209 TOMOTLEY
2210 TOMSHERRY
2211 TOMTY
2212 TONASKET
2213 TONATA
2214 TONCANA
2215 TONEY
2216 TONGUE RIVER
2217 TONIO
2218 TONKA
2219 TONKAWA
2220 TONKEY
2221 TONKIN
2222 TONKS
2223 TONOPAH
2224 TONOR
2225 TONOWEK
2226 TONRA
2227 TONSINA
2228 TONTI
2229 TONUCO
2230 TOOLESBORO
2231 TOOMES
2232 TOONE
2233 TOP
2234 TOPEKI
2235 TOPEMAN
2236 TOPETAUL
2237 TOPIA
2238 TOPLIFF
2239 TOPONCE
2240 TOPPENISH
2241 TOPSEY
2242 TOQUERVILLE
2243 TOQUI
2244 TOQUOP
2245 TORBOY
2246 TORCHLIGHT
2247 TORDIA
2248 TOREX
2249 TORHUNTA
2250 TORNEY
2251 TORNILLO
2252 TORNING
2253 TORODA
2254 TORONTO
2255 TORPEDO LAKE
2256 TORREON
2257 TORRES
2258 TORRO
2259 TORRY
2260 TORSIDO
2261 TORTUGAS
2262 TOSCA
2263 TOSTON
2264 TOTAVI
2265 TOTELAKE
2266 TOTEM
2267 TOTIER
2268 TOTO
2269 TOTTEN
2270 TOUCHET
2271 TOUHEY
2272 TOULON
2273 TOURNQUIST
2274 TOURS
2275 TOUTLE
2276 TOVAR
2277 TOWAVE
2278 TOWHEE
2279 TOWNER
2280 TOWNLEY
2281 TOWNSEND
2282 TOWOSAHGY
2283 TOXAWAY
2284 TOY
2285 TOYAH
2286 TOYKE
2287 TOYUSKA
2288 TOZE
2289 TRABUCO
2290 TRACHUTE
2291 TRACK
2292 TRACOSA
2293 TRACY
2294 TRADEDOLLAR
2295 TRAER
2296 TRAG
2297 TRAHAM
2298 TRAIL
2299 TRAILCREEK
2300 TRAMPAS
2301 TRAMWAY
2302 TRANQUILAR
2303 TRANSYLVANIA
2304 TRAPPER
2305 TRAPPIST
2306 TRAPPS
2307 TRASK
2308 TRAVELERS
2309 TRAVER
2310 TRAVERTINE
2311 TRAVESSILLA
2312 TRAVIS
2313 TRAWICK
2314 TRAY
2315 TREADWAY
2316 TREATY
2317 TREBLE
2318 TREBLOC
2319 TREBOR
2320 TREEKOR
2321 TREGO
2322 TREHARNE
2323 TRELK
2324 TRELONA
2325 TREMANT
2326 TREMBLES
2327 TREMONA
2328 TREMPE
2329 TREMPEALEAU
2330 TRENARY
2331 TRENT
2332 TRENTON
2333 TREON
2334 TREP
2335 TRES HERMANOS
2336 TRESANO
2337 TRESED
2338 TRESTLE
2339 TREVINO
2340 TREVLAC
2341 TREY
2342 TRIAMI
2343 TRIANGLE
2344 TRIBBEY
2345 TRICON
2346 TRID
2347 TRIDELL
2348 TRIGGER
2349 TRIGO
2350 TRIMAD
2351 TRIMBLE
2352 TRIMMER
2353 TRINITY
2354 TRIO
2355 TRIOMAS
2356 TRIPIT
2357 TRIPLEN
2358 TRIPOLI
2359 TRIPP
2360 TRISTAN
2361 TRITON
2362 TRIX
2363 TROCKEN
2364 TROJAN
2365 TROKEN
2366 TROMP
2367 TRONSEN
2368 TROOK
2369 TROPAL
2370 TROPIC
2371 TROPOHEMISTS
2372 TROSI
2373 TROSKY
2374 TROUGHS
2375 TROUP
2376 TROUT CREEK
2377 TROUT RIVER
2378 TROUTDALE
2379 TROUTER
2380 TROUTVILLE

2381 TROXEL
2382 TRUBLE
2383 TRUCE
2384 TRUCHOT
2385 TRUCKEE
2386 TRUCKTON
2387 TRUDAU
2388 TRUDE
2389 TRUEFISSURE
2390 TRUESDALE
2391 TRULON
2392 TRUMAN
2393 TRUMBULL
2394 TRUMP
2395 TRUNK
2396 TRYON
2397 TSCHICOMA
2398 TUB
2399 TUBAC
2400 TUCANNON
2401 TUCKAHOE
2402 TUCKER
2403 TUCKERMAN
2404 TUCSON
2405 TUCUMCARI
2406 TUFFIT
2407 TUFFO
2408 TUGHILL
2409 TUJUNGA
2410 TUKEY
2411 TUKUHNIK
2412 TUKWILA
2413 TULA
2414 TULANA
2415 TULARE
2416 TULARGO
2417 TULAROSA
2418 TULASE
2419 TULELAKE
2420 TULIA
2421 TULIK
2422 TULLAHASSEE
2423 TULLER
2424 TULLOCK
2425 TULLY
2426 TULOSO
2427 TUMAC
2428 TUMALO
2429 TUNBRIDGE
2430 TUNICA
2431 TUNIS
2432 TUNKHANNOCK
2433 TUOMI
2434 TUPELO
2435 TUPUKNUK
2436 TUQUE
2437 TURBEVILLE
2438 TURBOTVILLE
2439 TURBYFILL
2440 TURK
2441 TURKEYSPRINGS
2442 TURLEY
2443 TURLIN
2444 TURLOCK
2445 TURNBACK
2446 TURNBULL
2447 TURNER
2448 TURNERCREST
2449 TURNERVILLE
2450 TURNEY
2451 TURRAH
2452 TURRET
2453 TURRIA
2454 TURSON
2455 TUSAYAN
2456 TUSCAN
2457 TUSCARAWAS
2458 TUSCAWILLA
2459 TUSCOLA
2460 TUSCOSSO
2461 TUSCUMBIA
2462 TUSEL
2463 TUSK
2464 TUSKAHOMA
2465 TUSKEEGO
2466 TUSLER
2467 TUSQUITEE
2468 TUSSY
2469 TUSTIN
2470 TUSTUMENA
2471 TUTE
2472 TUTHILL
2473 TUTNI
2474 TUTWILER
2475 TUWEEP
2476 TWEBA
2477 TWEEDY
2478 TWEENER
2479 TWICK
2480 TWIG
2481 TWILIGHT
2482 TWIN CREEK
2483 TWINING
2484 TWINSI
2485 TWISSELMAN
2486 TWOMILE
2487 TWOTOP
2488 TYBO
2489 TYEE
2490 TYGART
2491 TYGH
2492 TYLER
2493 TYNDALL
2494 TYNER
2495 TYPIC ARIGIAQUOL
2496 TYPIC CRYAQUOLLS
2497 TYPIC CRYOBORALF
2498 TYPIC CRYOCHREPT
2499 TYPIC CRYORTHENT
2500 TYPIC CRYORTHODS
2501 TYPIC DYSTRANDEP
2502 TYPIC EUTROBORAL
2503 TYPIC FLUVAQUENT
2504 TYPIC HAPLAQUOLL
2505 TYPIC HYDRANDEPT
2506 TYPIC HYDRAQUENT
2507 TYPIC NADURARGID
2508 TYPIC TORRIFLUVE
2509 TYPIC TORRIORTHE
2510 TYPIC TROPOFOLIS
2511 TYPIC TROPORTHEN
2512 TYPIC USTIFLUVEN
2513 TYPIC USTORTHENT
2514 TYPIC USTROPEPTS
2515 TYPIC VITRANDEPT
2516 TYPIC XEROFLUVEN
2517 TYPIC XERORTHENT
2518 TYRE
2519 TYRONE
2520 TYSON

SOIL06.DAT

1 UANA
2 UBANK
3 UBAR
4 UBIK
5 UBLY
6 UCHEE
7 UCOLO
8 UCOPIA
9 UDAHO
10 UDEL
11 UDELOPE
12 UDIC USTOCHREPTS
13 UDOLPHO
14 UDORTHENTS
15 UDORTHENTS,
16 UDORTHENTS, LOAM
17 UFFENS
18 UGAK
19 UHALDI
20 UHL
21 UHLAND
22 UHLIG
23 UHLORN
24 UINTA
25 UKIAH
26 ULA
27 ULEN
28 ULIDA
29 ULM
30 ULRIC
31 ULRICHER
32 ULTRA
33 ULUPALAKUA
34 ULY
35 ULYSSES
36 UMA
37 UMAPINE
38 UMATILLA
39 UMBARG
40 UMBERLAND
41 UMBL
42 UMBRIC DYSTROCHR
43 UMIAT
44 UMIKOA
45 UMIL
46 UMPA
47 UMPCOOS
48 UMPUMP
49 UNA
50 UNADILLA
51 UNAKA
52 UNAKWIK
53 UNAWEEP
54 UNCAS
55 UNCOMPAHGRE
56 UNDERWOOD
57 UNDUSK
58 UNGERS
59 UNICOI
60 UNION
61 UNIONTOWN
62 UNIONVILLE
63 UNISON
64 UNIUS
65 UNIVEGA
66 UNLIC
67 UNSEL
68 UNSON
69 UPDIKE
70 UPSATA
71 UPSHUR
72 UPSON
73 UPSPRING
74 UPTMOR
75 UPTON
76 UPVILE
77 URACCA
78 URBAN LAND
79 URBANA
80 URBO
81 UREAL
82 URICH
83 URIPNES
84 URLAND
85 URNE
86 URNESS
87 URSA
88 URSINE
89 URTAH
90 URWIL
91 USAL
92 USHAR
93 USINE
94 USK
95 USTIC TORRIFLUVE
96 USTIC TORRIORTHE
97 USTIC TORRIPSAMM
98 USTIFLUVENTS, BR
99 USTOCHREPTS
100 USTOLLIC CALCIOR
101 USTOLLIC HAPLARG
102 USTOLLIC PALEORT
103 USTOLLS
104 USTORTENTS
105 USTORTHENTS,STEE
106 UTABA
107 UTALINE
108 UTE
109 UTICA
110 UTLEY
111 UTSO
112 UTUADO
113 UVADA
114 UVALDE
115 UVI
116 UWALA
117 UWHARRIE
118 VABEM
119 VABUS
120 VACHERIE
121 VADER
122 VADNAIS
123 VADO
124 VAEDA
125 VAIDEN
126 VAIDEN CALCAREOU
127 VAILTON
128 VAIVA
129 VALBY
130 VALCO
131 VALCREST
132 VALDEZ
133 VALDOSTA
134 VALE
135 VALENCIA
136 VALENT
137 VALENTINE
138 VALERA
139 VALHALLA
140 VALKARIA
141 VALLAN
142 VALLE
143 VALLECITOS
144 VALLEONO
145 VALLERS
146 VALLEYCITY
147 VALMAR
148 VALMONT
149 VALMY
150 VALNOR
151 VALOIS
152 VALSETZ
153 VALTO
154 VALTON
155 VALVERDE
156 VAMER
157 VAMONT
158 VAMP
159 VAN DUSEN
160 VAN HORN
161 VAN HORN, WET
162 VAN NOSTERN
163 VAN WAGONER
164 VANAJO
165 VANANDA
166 VANCE
167 VANDA
168 VANDALIA
169 VANDAMME
170 VANDAMORE
171 VANDERGRIFT
172 VANDERHOFF
173 VANDERLIP
174 VANEPPS
175 VANET
176 VANG
177 VANGUARD
178 VANMETER
179 VANNI
180 VANNOY
181 VANOCKER
182 VANOSS
183 VANPETTEN
184 VANSON
185 VANSTEL
186 VANTAGE
187 VANWYPER
188 VANZANDT
189 VAQUERO
190 VARCO
191 VARDEN
192 VARELUM
193 VARGUS
194 VARICK
195 VARINA
196 VARNA
197 VARNEY
198 VARRO
199 VARYSBURG
200 VASA
201 VASHTI
202 VASQUEZ
203 VASSALBORO
204 VASSAR
205 VASSETT
206 VASTINE
207 VAUCLUSE
208 VAUGHAN
209 VAUGHNSVILLE
210 VAY
211 VAYAS
212 VEAL
213 VEATCH
214 VEAZIE
215 VEBAR
216 VECONT
217 VEEDUM
218 VEET
219 VEGA
220 VEGA ALTA
221 VEGA BAJA
222 VEKOL
223 VELASCO
224 VELDA
225 VELDKAMP
226 VELMA
227 VELOW
228 VELVA
229 VENA
230 VENABLE
231 VENANGO
232 VENATOR
233 VENETA
234 VENEZIA
235 VENICE
236 VENLO
237 VENTRIS
238 VENTURE
239 VENUM
240 VENUS
241 VERBOORT
242 VERDE
243 VERDEL
244 VERDICO
245 VERDIGRIS
246 VERDUN
247 VERGAS
248 VERGENNES
249 VERHALEN
250 VERICK
251 VERITAS
252 VERLAND
253 VERLOT
254 VERMEJO
255 VERMILAC
256 VERMILLON
257 VERMISA
258 VERNAL
259 VERNALIS
260 VERNIA
261 VERNIGOR
262 VERNON
263 VERNONIA
264 VERO
265 VERSHIRE
266 VERSON
267 VERTEL
268 VERTREES
269 VES
270 VESEY
271 VESPER
272 VESSER
273 VESSILLA
274 VESTA
275 VESTABURG
276 VESTON
277 VETA
278 VETAL
279 VETEADO
280 VEYO

281 VIA
282 VIAN
283 VIBLE
284 VIBO
285 VIBORAS
286 VIBORG
287 VICEE
288 VICK
289 VICKERY
290 VICKING
291 VICKSBURG
292 VICTINE
293 VICTOR
294 VICTORIA
295 VICTORVILLE
296 VICTORY
297 VICU
298 VIDA
299 VIDAURI
300 VIDRINE
301 VIEJA
302 VIENNA
303 VIEQUES
304 VIGAR
305 VIGIA
306 VIGNOLO
307 VIGO
308 VIGUS
309 VIKING
310 VIL
311 VILAS
312 VILLA
313 VILLA GROVE
314 VILLEGREEN
315 VILLY
316 VILOT
317 VIMVILLE
318 VINA
319 VINCENNES
320 VINCENT
321 VINCOM
322 VINDICATOR
323 VINEGARROON
324 VINEYARD
325 VINGO
326 VINING
327 VININI
328 VINITA
329 VINLAND
330 VINSAD
331 VINSON
332 VINT
333 VINTAS
334 VINTON
335 VIPONT
336 VIRATON
337 VIRDEN
338 VIRGELLE
339 VIRGIL
340 VIRGIN PEAK
341 VIRGIN RIVER
342 VIRKULA
343 VIRLE
344 VIRTUE
345 VISTA
346 VITALE
347 VIUDA
348 VIVES
349 VIVI
350 VIZCAINO
351 VLASATY
352 VOATS
353 VOCA
354 VODERMAIER
355 VOIGHT
356 VOLADORA
357 VOLASH
358 VOLBORG
359 VOLCO
360 VOLENTE
361 VOLINIA
362 VOLKMAR
363 VOLNEY
364 VOLPERIE
365 VOLTA
366 VOLTAIRE
367 VOLUSIA
368 VONA
369 VONALEE
370 VOORHIES
371 VORE
372 VOSBURG
373 VOSS
374 VOSSET
375 VULCAN
376 VYLACH
377 WAAS
378 WABASH
379 WABASHA
380 WABASSO
381 WABBASEKA
382 WABEK
383 WABEN
384 WABUSKA
385 WACA
386 WACAHOOTA
387 WACOTA
388 WACOUSTA
389 WADAMS
390 WADDOUPS
391 WADELL
392 WADENA
393 WADENILL
394 WADER
395 WADESPRING
396 WADMALAW
397 WADSWORTH
398 WAGES
399 WAGNER
400 WAGONBOX
401 WAGONTIRE
402 WAGRAM
403 WAHA
404 WAHATOYA
405 WAHEE
406 WAHGUYHE
407 WAHIAWA
408 WAHIKULI
409 WAHKEENA
410 WAHLUKE
411 WAHOO
412 WAHPETON
413 WAHREKDAM
414 WAHTIGUP
415 WAHTUM
416 WAIAHA
417 WAIAKOA
418 WAIALEALE
419 WAIALUA
420 WAIAWA
421 WAIHUNA
422 WAIKALOA
423 WAIKANE
424 WAIKAPU
425 WAIKOMO
426 WAILUKU
427 WAIMEA
428 WAINEE
429 WAINOLA
430 WAIPAHU
431 WAISKA
432 WAITS
433 WAKE
434 WAKEEN
435 WAKEFIELD
436 WAKELAND
437 WAKITA
438 WAKONDA
439 WAKULLA
440 WALCOTT
441 WALDBILLIG
442 WALDECK
443 WALDEN
444 WALDO
445 WALDORF
446 WALDPORT
447 WALDRON
448 WALDROUP
449 WALES
450 WALFORD
451 WALHALLA
452 WALKE
453 WALKNOLLS
454 WALKON
455 WALL
456 WALLA WALLA
457 WALLACE
458 WALLEN
459 WALLER
460 WALLINGTON
461 WALLKILL
462 WALLROCK
463 WALLSBURG
464 WALLSON
465 WALLUSKI
466 WALNETT
467 WALONG
468 WALPOLE
469 WALREES
470 WALSH
471 WALSTEAD
472 WALTERS
473 WALTI
474 WALUM
475 WALVAN
476 WALVILLE
477 WAMBA
478 WAMBOLT
479 WAMDUSKA
480 WAMIC
481 WAMPOO
482 WAMPSVILLE
483 WANBLEE
484 WANDA
485 WANDO
486 WANETTA
487 WANILLA
488 WANN
489 WANOGA
490 WANSER
491 WAPAL
492 WAPATO
493 WAPELLO
494 WAPI
495 WAPINITIA
496 WAPPING
497 WAPPINGER
498 WAPPO
499 WAPSHILLA
500 WAPSIE
501 WAPTUS
502 WARBA
503 WARDBORO
504 WARDELL
505 WARDEN
506 WARDENOT
507 WARDWELL
508 WARE
509 WAREHAM
510 WARM SPRINGS
511 WARMAN
512 WARNEKE
513 WARNERS
514 WARNOCK
515 WARRENTON
516 WARSAW
517 WARSING
518 WARWICK
519 WASA
520 WASATCH
521 WASCO
522 WASDA
523 WASEPI
524 WASHBURN
525 WASHINGTON
526 WASHOE
527 WASHOUGAL
528 WASHTENAW
529 WASILLA
530 WASIOJA
531 WASKISH
532 WASKOM
533 WASPO
534 WASSAIC
535 WASSIT
536 WATAB
537 WATAMA
538 WATAUGA
539 WATCHABOB
540 WATCHAUG
541 WATCHUNG
542 WATERBURY
543 WATERCANYON
544 WATEREE
545 WATERINO
546 WATERMAN
547 WATERTOWN
548 WATKINS
549 WATKINS RIDGE
550 WATO
551 WATONGA
552 WATOOPAH
553 WATROUS
554 WATSEKA
555 WATSON
556 WATSONIA
557 WATSONVILLE
558 WATT
559 WATTON
560 WATUSI
561 WAUBAY
562 WAUBEEK
563 WAUBERG
564 WAUBONSIE
565 WAUCEDAH
566 WAUCHULA
567 WAUCOBA
568 WAUCOMA
569 WAUCONDA
570 WAUKEE
571 WAUKEGAN
572 WAUKENA
573 WAUKON
574 WAULD
575 WAUMBEK
576 WAUPACA
577 WAUPECAN
578 WAURIKA
579 WAUSEON
580 WAUTOMA

581 WAVELAND
582 WAVERLY
583 WAWASEE
584 WAWINA
585 WAX
586 WAXPOOL
587 WAYBE
588 WAYCUP
589 WAYDEN
590 WAYLAND
591 WAYMOR
592 WAYNECO
593 WAYNESBORO
594 WAYNETOWN
595 WEA
596 WEASH
597 WEATHERFORD
598 WEAVER
599 WEBB
600 WEBBRIDGE
601 WEBBTOWN
602 WEBER
603 WEBILE
604 WEBSTER
605 WEDEKIND
606 WEDERTZ
607 WEDGE
608 WEDLAR
609 WEDOWEE
610 WEED
611 WEEDING
612 WEEDMARK
613 WEEKIWACHEE
614 WEEKS
615 WEEKSVILLE
616 WEENA
617 WEEPAH
618 WEESATCHE
619 WEGA
620 WEHADKEE
621 WEIGANG
622 WEIGLE
623 WEIKERT
624 WEIMER
625 WEINBACH
626 WEINGART
627 WEINGARTEN
628 WEIR
629 WEIRMAN
630 WEISBURG
631 WEISER
632 WEISHAUPT
633 WEISSENFELS
634 WEITCHPEC
635 WELAKA
636 WELBY
637 WELCH
638 WELCHLAND
639 WELCOME
640 WELD
641 WELDA
642 WELLER
643 WELLINGTON
644 WELLMAN
645 WELLS
646 WELLSBORO
647 WELLSED
648 WELLSTON
649 WELLSVILLE
650 WELLTON
651 WELOY
652 WELRING
653 WELTER
654 WEMPLE
655 WENAS
656 WENATCHEE
657 WENDANE
658 WENDOVER
659 WENDTE
660 WENONA
661 WENTWORTH
662 WENZEL
663 WEOGUFKA
664 WERELD
665 WERLOG
666 WERNER
667 WERNOCK
668 WESCONNETT
669 WESKA
670 WESLEY
671 WESO
672 WESSEL
673 WESTBROOK
674 WESTBURY
675 WESTBUTTE
676 WESTCAMP
677 WESTCREEK
678 WESTERVILLE
679 WESTFORK
680 WESTHAVEN
681 WESTINDIAN
682 WESTLAKE
683 WESTLAND
684 WESTMORE
685 WESTMORELAND
686 WESTON
687 WESTOVER
688 WESTPHALIA
689 WESTPLAIN
690 WESTPORT
691 WESTRACO
692 WESTSHORE
693 WESTVACO
694 WESTVIEW
695 WESTVILLE
696 WESTWEGO
697 WESWOOD
698 WETA
699 WETHERSFIELD
700 WETHEY
701 WETMORE
702 WETSAW
703 WETTERHORN
704 WETZEL
705 WEVERTON
706 WEWELA
707 WEWOKA
708 WEYERS
709 WEYMOUTH
710 WHAKANA
711 WHALAN
712 WHALEN
713 WHALEY
714 WHARTON
715 WHATCOM
716 WHATELY
717 WHEATLEY
718 WHEATRIDGE
719 WHEATVILLE
720 WHEELER
721 WHEELERVILLE
722 WHEELING
723 WHEELON
724 WHETROCK
725 WHETSTONE
726 WHICHMAN
727 WHIDBEY
728 WHIPPANY
729 WHIPSTOCK
730 WHIRLO
731 WHISPERING
732 WHISTLE
733 WHIT
734 WHITAKER
735 WHITE HOUSE
736 WHITE STORE
737 WHITE SWAN
738 WHITECAP
739 WHITECLOUD
740 WHITECOW
741 WHITECROSS
742 WHITEFISH
743 WHITEFORD
744 WHITEHALL
745 WHITEHILLS
746 WHITEHORN
747 WHITEHORSE
748 WHITEKNOB
749 WHITELAKE
750 WHITEMAN
751 WHITEPEAK
752 WHITEROCK
753 WHITESBORO
754 WHITESBURG
755 WHITESON
756 WHITEWATER
757 WHITEWOLF
758 WHITEWOOD
759 WHITEWRIGHT
760 WHITLEY
761 WHITLOCK
762 WHITMAN
763 WHITNEY
764 WHITORE
765 WHITSOL
766 WHITSON
767 WHITTIER
768 WHITWELL
769 WHOBREY
770 WHOLAN
771 WIBAUX
772 WICHITA
773 WICHUP
774 WICKAHONEY
775 WICKENBURG
776 WICKERSHAM
777 WICKETT
778 WICKHAM
779 WICKIUP
780 WICKSBURG
781 WIDEMAN
782 WIDEN
783 WIDTSOE
784 WIEHL
785 WIELAND
786 WIERGATE
787 WIGGLETON
788 WIGTON
789 WILAHA
790 WILBANKS
791 WILBRAHAM
792 WILBUR
793 WILBURTON
794 WILCO
795 WILCOX
796 WILCOXSON
797 WILDALE
798 WILDCAT
799 WILDERNESS
800 WILDHORSE
801 WILDORS
802 WILDWOOD
803 WILE
804 WILEY
805 WILHITE
806 WILHOIT
807 WILKES
808 WILKESON
809 WILKINS
810 WILL
811 WILLABY
812 WILLACY
813 WILLAKENZIE
814 WILLAMAR
815 WILLAMETTE
816 WILLANCH
817 WILLAPA
818 WILLARD
819 WILLETTE
820 WILLHILL
821 WILLHO
822 WILLIAMS
823 WILLIAMSBURG
824 WILLIAMSON
825 WILLIAMSTOWN
826 WILLIAMSVILLE
827 WILLIMAN
828 WILLIS
829 WILLISTON
830 WILLOW CREEK
831 WILLOWDALE
832 WILLOWMAN
833 WILLOWS
834 WILLWOOD
835 WILMER
836 WILMINGTON
837 WILMONTON
838 WILPAR
839 WILPOINT
840 WILSHIRE
841 WILSON
842 WILSONVILLE
843 WILSOR
844 WILTON
845 WINADA
846 WINBERRY
847 WINCHESTER
848 WINCHUCK
849 WIND RIVER
850 WIND RIVER, GRAV
851 WINDER
852 WINDFALL
853 WINDHAM
854 WINDICREEK
855 WINDMILL
856 WINDSOR
857 WINDTHORST
858 WINDWHISTLE
859 WINDY
860 WINEG
861 WINEMA
862 WINETTI
863 WINEVADA
864 WINFALL
865 WINFIELD
866 WING
867 WINGATE
868 WINGER
869 WINGINAW
870 WINGVILLE
871 WINIFRED
872 WINK
873 WINKEL
874 WINKLEMAN
875 WINKLER
876 WINLER
877 WINLO
878 WINN
879 WINNEBAGO
880 WINNECONNE

881 WINNECOOK
882 WINNEMUCCA
883 WINNESHIEK
884 WINNETT
885 WINNSBORO
886 WINONA
887 WINOOSKI
888 WINOPEE
889 WINRIDGE
890 WINSHIP
891 WINSPECT
892 WINSTON
893 WINT
894 WINTERFIELD
895 WINTERHAVEN
896 WINTERIDGE
897 WINTERS
898 WINTERSBURG
899 WINTERSET
900 WINTHROP
901 WINTLEY
902 WINTON
903 WINTONER
904 WINU
905 WINZ
906 WIOTA
907 WIRT
908 WISCOW
909 WISE
910 WISEMAN
911 WISENOR
912 WISHARD
913 WISHBONE
914 WISHEYLU
915 WISHKAH
916 WISKAN
917 WISNER
918 WISTER
919 WITBECK
920 WITEFELS
921 WITHAM
922 WITHEE
923 WITHERBEE
924 WITHERELL
925 WITHERS
926 WITT
927 WITTEN
928 WITTENBERG
929 WITZEL
930 WIX
931 WIXOM
932 WOCKLEY
933 WODA
934 WODEN
935 WODSKOW
936 WOHLY
937 WOLCO
938 WOLCOTT
939 WOLDALE
940 WOLF
941 WOLF POINT
942 WOLFCREEK
943 WOLFESON
944 WOLFPEN
945 WOLFTEVER
946 WOLLARD
947 WOLLENT
948 WOLOT
949 WOLVERINE
950 WOO
951 WOOD RIVER
952 WOODBECK
953 WOODBINE
954 WOODBRIDGE
955 WOODBURN
956 WOODBURY
957 WOODCOCK
958 WOODFORD
959 WOODGULCH
960 WOODHALL
961 WOODHURST
962 WOODIN
963 WOODINGTON
964 WOODINVILLE
965 WOODLAWN
966 WOODLY
967 WOODLYN
968 WOODMANSIE
969 WOODMERE
970 WOODMONT
971 WOODPASS
972 WOODROCK
973 WOODROW
974 WOODS CROSS
975 WOODSEYE
976 WOODSFIELD
977 WOODSIDE
978 WOODSON
979 WOODSTOCK
980 WOODSTOWN
981 WOODTELL
982 WOODVILLE
983 WOODWARD
984 WOOLPER
985 WOOLSEY
986 WOOLSTALF
987 WOOLSTED
988 WOONSOCKET
989 WOOSLEY
990 WOOSTER
991 WORCESTER
992 WORDEN
993 WORF
994 WORFKA
995 WORFMAN
996 WORK
997 WORL
998 WORLAND
999 WORLEY
1000 WORMSER
1001 WOROCK
1002 WORSHAM
1003 WORTH
1004 WORTHEN
1005 WORTHING
1006 WORTMAN
1007 WOVOKA
1008 WRANGO
1009 WREDAH
1010 WRENCOE
1011 WRENMAN
1012 WRENTHAM
1013 WRIGHT
1014 WRIGHTMAN
1015 WRIGHTSBORO
1016 WRIGHTSVILLE
1017 WRIGHTWOOD
1018 WUKOKI
1019 WUKSI
1020 WULFERT
1021 WUNJEY
1022 WUPATKI
1023 WURNO
1024 WURSTEN
1025 WURTSBORO
1026 WYALUSING
1027 WYANDOTTE
1028 WYARD
1029 WYARNO
1030 WYATT
1031 WYE
1032 WYEAST
1033 WYETH
1034 WYEVILLE
1035 WYICK
1036 WYKEHAM
1037 WYKOFF
1038 WYMAN
1039 WYMORE
1040 WYNDMERE
1041 WYNN
1042 WYNNVILLE
1043 WYNONA
1044 WYNOOSE
1045 WYOCENA
1046 WYOMING
1047 WYRENE
1048 WYSOCKING
1049 XANA
1050 XAVIER
1051 XENIA
1052 XENO
1053 XERERTIC TORRIOR
1054 XERIC TORRIFLUVE
1055 XERIC TORRIORTHE
1056 XEROCHREPTS
1057 XEROFLUVENT
1058 XEROFLUVENTS
1059 XEROLLIC CAMBORT
1060 XEROLLIC HAPLARG
1061 XEROPSAMMENTS
1062 XERORTHENTS
1063 XERTA
1064 XERUMBREPTS
1065 XINE
1066 XMAN
1067 YACOLT
1068 YAGO
1069 YAHARA
1070 YAHNE
1071 YAHOLA
1072 YAINAX
1073 YAKI
1074 YAKIMA
1075 YAKUS
1076 YAKUTAT
1077 YALELAKE
1078 YALESVILLE
1079 YALLANI
1080 YALMER
1081 YAMAC
1082 YAMHILL
1083 YAMO
1084 YAMPA
1085 YAMSAY
1086 YANA
1087 YANCY
1088 YANKEE
1089 YANKTON
1090 YANUSH
1091 YAP
1092 YAPOAH
1093 YAQUINA
1094 YARCO
1095 YARDLEY
1096 YARTS
1097 YATAHONEY
1098 YATES
1099 YAUCO
1100 YAUHANNAH
1101 YAUPON
1102 YAWDIM
1103 YAWHEE
1104 YAWKEY
1105 YAXON
1106 YEARY
1107 YEATES HOLLOW
1108 YECROSS
1109 YEDLICK
1110 YEGEN
1111 YELJACK
1112 YELLOWBAY
1113 YELLOWHOUND
1114 YELLOWROCK
1115 YELLOWSTONE
1116 YELM
1117 YEMASSEE
1118 YENCE
1119 YENLO
1120 YENRAB
1121 YEOMAN
1122 YEOPIM
1123 YERINGTON
1124 YERMO
1125 YESO
1126 YESUM
1127 YETTEM
1128 YETULL
1129 YIPOR
1130 YOBE
1131 YOCHUM
1132 YOCKEY
1133 YODER
1134 YODY
1135 YOHURT
1136 YOKAYO
1137 YOKOHL
1138 YOLLABOLLY
1139 YOLO
1140 YOLOGO
1141 YOMBA
1142 YOMONT
1143 YONCALLA
1144 YONGES
1145 YONNA
1146 YORBA
1147 YORK
1148 YORKTOWN
1149 YORKTREE
1150 YORKVILLE
1151 YOST
1152 YOUD
1153 YOUGA
1154 YOUJAY
1155 YOUMAN
1156 YOUNGSTON
1157 YOURAME
1158 YOUTLKUE
1159 YOVIMPA
1160 YPSI
1161 YRIBARREN
1162 YSIDORA
1163 YTURBIDE
1164 YTURRIA
1165 YUBA
1166 YUKO
1167 YUKON
1168 YULEE
1169 YUNES
1170 YUNQUE
1171 YURM
1172 YUTRUE
1173 YUVAS
1174 ZAAR
1175 ZABA
1176 ZACA
1177 ZACHARIAS
1178 ZACHARY
1179 ZADOG
1180 ZADVAR

1181 ZAFRA
1182 ZAGG
1183 ZAHILL
1184 ZAHL
1185 ZAIDY
1186 ZAKME
1187 ZALCO
1188 ZALDA
1189 ZALLA
1190 ZAMORA
1191 ZAMSCAN
1192 ZANE
1193 ZANEIS
1194 ZANESVILLE
1195 ZANGO
1196 ZAPATA
1197 ZAU
1198 ZAVALA
1199 ZAVCO
1200 ZAYANTE
1201 ZAZA
1202 ZEALE
1203 ZEB
1204 ZEBA
1205 ZECANYON
1206 ZEEBAR
1207 ZEEKA
1208 ZEESIX
1209 ZEGRO
1210 ZEIBRIGHT
1211 ZELL
1212 ZEN
1213 ZENDA
1214 ZENI
1215 ZENIFF
1216 ZENITH
1217 ZENKER
1218 ZENOD
1219 ZENOR
1220 ZEONA
1221 ZEORELY
1222 ZEPHAN
1223 ZEPHYR
1224 ZER
1225 ZERK
1226 ZERKER
1227 ZIEGENFUSS
1228 ZIEGLER
1229 ZIGWEID
1230 ZILABOY
1231 ZILLAH
1232 ZILLION
1233 ZILLMAN
1234 ZIMMERMAN
1235 ZINEB
1236 ZING
1237 ZINZER
1238 ZION
1239 ZIPP
1240 ZIRAM
1241 ZITA
1242 ZITTAU
1243 ZOAR
1244 ZOATE
1245 ZOE
1246 ZOESTA
1247 ZOHNER
1248 ZOLA
1249 ZOLFO
1250 ZOLLIN
1251 ZOLTAY
1252 ZOOK
1253 ZORRA
1254 ZORRAVISTA
1255 ZOYER
1256 ZUBER
1257 ZUFELT
1258 ZUKAN
1259 ZUMAN
1260 ZUMBRO
1261 ZUMWALT
1262 ZUNDELL
1263 ZUNHALL
1264 ZUNI
1265 ZURICH
1266 ZWICKER
1267 ZWIEFEL
1268 ZWINGLE
1269 ZYGORE
1270 ZYME
1271 ZYMER
1272 ZYNBAR
1273 ZYZYL
1274 ZYZZI
1275 ZYZZUG

Appendix VI

Topographic Maps

APPENDIX VI

Sources of Topographic Maps

U.S. Department of the Interior
Geological Survey

1. Eastern Region - Map Distribution
   U.S. Geological Survey
   1200 South Eads Street
   Arlington, VA 22202

2. Western Region - Map Distribution
   U.S. Geological Survey
   Box 25286 Denver Federal Center
   Denver, CO 80225

Appendix VII

Crop Data

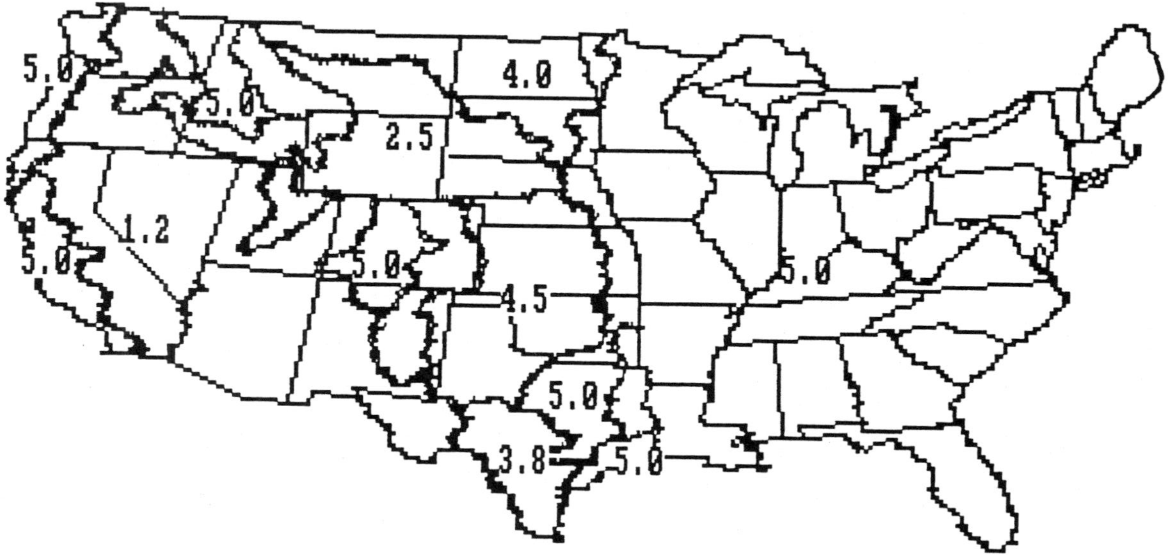

Appendix VII. **Maximum Leaf Area Index**

Appendix VIII

Input Data

APPENDIX VIII

Information Input and Format	Variables	Definitions	Units	Typical Range	Where Information is Obtained
1. Title 20A4	TITLE				User supplies
2. Program control codes 20I4	NBYR	Number years of runoff simulation		1-100	User specifies
	IYR	Beginning year of runoff simulation		1-2000	"
	LU	Number of subareas in basin		1-5	"
	IPD	Print code--0 = monthly, 1 = daily, 2 = yearly		0-2	"
	NSIM	Code for rainfall input--1 = measured, 2 = simulated single, 3 = simulated multiple		1-3	"
	MSIM	Code for temperature input--1 = measured temperature 2 = simulated single temperatures for entire basin, 3 = simulated temperature for each subbasin		0-2	"
	IGN	Number of times random number generator cycles before simulation begins		0-100	"
3. General data 10F8.3	DA	Basin area	km^2	1-5000	User specifies
	P2	Rainfall correction factor		0.5-1.5	"
	BFF	Baseflow factor		0-1	Table III.4
	BRT	Basin lag time	days	0-120	User specifies
	FFC	Initial soil water storage as a fraction of field capacity		0-1	"
4. Sub-basin centroid coordinates 10F8.3	XIJ	X-centroid coordinates of subbasins	km	0-200	User specifies
	YIJ	Y-coordinates of subbasins	km	0-200	"

Information Input and Format	Variables	Definitions	Units	Typical Range	Where Information is Obtained
5. General weather data 10F8.3	TP5	TP-40 10-yr frequency 0.5 h rainfall	mm	5.0-150	Table II.1
	TP6	TP-40 10-yr frequency 6.0 h rainfall	mm	25-200	"
	TP24	Number years record max 0.5 h rain		5-100	"
	YLT	Latitude of watershed	deg	-90-90	"
6. Temperature data 12F6.3	OBMX	Average monthly maximum air temperature	C	-10-42	Table II.1
	OBMN	Average monthly minimum air temperature	C	-30-30	"
	VOBMX	If MSIM = 3, average monthly maximum air temperature	C	-10-42	"
	VOBMN	If MSIM = 3, average monthly minimum air temperature	C	-30-30	"
	CVT	Coefficient of variation for monthly temperature		0-1	"
7. Solar radiation data 12F6.3	OBSL	Monthly average daily solar radiation	ly	20-750	Table II.1
8. Monthly rainfall data 12F6.3	WI	Monthly maximum 0.5 h rainfall	mm	0-125	Table II.1
9. Precipitation parameters 12F8.3	PRW(1)	Monthly probability of wet day after a dry day	---	.001-.95	Table II.1
	PRW(2)	Monthly probability of a wet day after a wet day	---	.01-.95	"
	WLV	Days of precipitation in a month	---	0-31	User specifies
	RST(1)	Monthly mean of daily precipitation	mm	0-500	Table II.1
	RST(2)	Monthly standard deviation of daily precipitation	mm	.05-200	"
	RST(3)	Monthly skew coefficient of daily precipitation	---	1-7	"

Information Input and Format	Variables	Definitions	Units	Typical Range	Where Information is Obtained
10. Basin data 10F8.3	FLU	Fraction of basin in each subarea		0-1	User specifies
	CN2	SCS runoff curve number--condition II		5-95	Table III.1
	SALB	Soil albedo		.05-.20	User specifies
	SNO	Initial water content of snow	mm	0-150	User specifies
	CHL1	Channel length from most distant point to subbasin outlet	km	.05-200	Appendix VI
	CHS	Average channel slope	m/m	.0001-10	"
	CHK1	Effective hydraulic conductivity in channel alluvium	mm	0-150	Table III.3
	CHW1	Average main channel width	m	1-1000	User specifies
	CHN	Channel n value		.01-.3	Table III.2
	OVN	Overland flow N value		.01-.3	"
	RT	Return flow travel time	days	0-180	User specifies
	CSS	Sediment concentration in return flow	ppm	100-1500	"
	ECP	USLE erosion control practice factor P		.1-1.0	Table IV.1
	SL	Average slope length for each sub-basin	m	10-150	Appendix VI
	STP	Average slope steepness for each sub-basin	m/m	.0001-.60	"
11. Routing data (sub-basin to basin outlet) 10F8.3	CHW2	Average channel width	m	1-1000	Appendix VI
	CHD	Average channel depth	m	.5-30	"
	CHSS	Channel slope	m/m	.0001-10	"
	CHL2	Channel length	km	.05-200	"
	CHNN	Channel n value		.01-.3	Table III.2
	CHK2	Effective hydraulic conductivity in channel alluvium	mm/hr	0-150	Table III.3
	CHXK	USLE soil factor K for channel	---	.05-.60	Appendix V
	CHC	USLE soil factor C for channel	---	.001-1	"

Information Input and Format	Variables	Definitions	Units	Typical Range	Where Information is Obtained
12. Pond data 10F8.3	FP	Fraction of each subbasin that flows into ponds		0-1	User supplies
	SAX	Total surface area of all ponds in each subbasin	ha	0-1000	"
	VMX	Runoff volume from pond catchment area required to fill empty ponds	mm	0-100	"
	VM	Initial pond volumes	mm	0-100	"
	CS	Initial sediment concentration	ppm	0-5000	"
	CFP	Normal sediment concentration	ppm	0-5000	"
	HC	Hydraulic conductivity of pond bottoms	mm/hr	0-1	"
13. Reservoir data 10F8.3	FR	Fraction of each subbasin that flows into reservoirs		0-1	User supplies
	SAF	Total reservoir surface area at emergency spillway	ha	0-3000	"
	VRF	Runoff volume from reservoir catchment area required to fill emergency spillway	mm	0-300	"
	SAS	Total reservoir surface area at principle spillway	ha	0-1000	"
	VRS	Runoff required to fill to principle spillway	mm	0-100	"
	VR	Initial reservoir volumes	mm	0-100	"
	RRR	Average principle spillway release rate	$m^3/s/km^2$	0-1	"
	CSR	Initial sediment concentration in reservoirs	ppm	0-5000	"
	CFR	Normal sediment concentration in reservoirs	ppm	0-5000	"
	HCR	Hydraulic conductivity of reservoir bottoms	mm/h	0-1	"

Information Input and Format	Variables	Definitions	Units	Typical Range	Where Information is Obtained
14. Soil data 20I4 and 10F8.3	NSO	Number of soils	---	1-5	User specifies
	MS	Soil that corresponds to given subbasin	---	1-5	"
	NS1	Number of soil layer from each soil	---	2-10	"
	EK	USLE soils factor K	---	.05-.60	Appendix V
	SIL	Pass #200 sieve	%	0-100	"
	Z	Depth to bottom of layers	mm	10-3500	"
	POR	Bulk density	t/m^3	.05-2.6	"
	T4	Available water capacity	mm/mm	0-1.0	"
	SC	Saturated conductivity	mm/h	0.200	"
	CLA	Clay content	%	0-100	"
	SIL	Pass #200 sieve	%	0-100	
	PSZ	Particle size distribution	fraction	0-1	"
15. Crop data 20I4 and 10F8.3	MO	Month of planting	mos	1-12	User specifies
	IDA	Day of month of planting	days	1-31	"
	MOH	Month of harvest	mos	1-12	"
	IDH	Day of month of harvest	days	1-31	"
	IRD	Vegetation--1 = annual (crops), 2 = perennial (forest, range, meadow, etc.)		1-2	"
	ITIL	Code for tillage operation--1 = fall plow, 2 = spring plow, 3 = conservation tillage, 4 = zero tillage		1-4	"
	CVM	Average annual C factor for subbasin		.001-.5	User specifies
	BLAI	Maximum LAI for sub-basin		.5-7	"
	ZMX	Maximum rooting depth	mm	100-3000	"

Information Input and Format	Variables	Definitions	Units	Typical Range	Where Information is Obtained
16. Irrigation data I4, 2F8.2	IRR	Irrigation code--1 = irrigate, 0 = no irrigation	---	0-1	User specifies
	WSF	Value of water stress factor when irrigation begins	---	0-.95	"
	EFI	Irrigation runoff rates	---	0-1.0	"
17. Daily rainfall and temperature data 5X, 15F5.1		Daily precipitation for each subbasin	mm	0-250	User supplies
		Daily maximum and minimum temperature for each subbasin	C	-50-54	"

Appendix IX

Input Data Forms

This page was intentionally left blank.

*** SWRRB DATA ASSEMBLY FORM 1 ***

1. **TITLE:** ______________________________
____
____

2. **PROGRAM CONTROL CODES**

Number of years of runoff simulation ________

Beginning year of runoff simulation ________

Number of subareas in basin ________

Printout frequency

Monthly (0) ________ Daily (1) ________ Annual (2) ________

Will the rainfall be:

Read in single rain gage for entire basin (1) ________

Simulated single rain gage for entire basin (2) ________

Read in one rain gage for each sub-basin (3) ________

Simulated for multiple rain gages (4) ________

Will the max & min temperatures be:

Read in single max & min for entire basin (1) __________

Simulated single max & min for entire basin (2) __________

Read in max & min for each sub-basin (3) __________

Simulated for each sub-basin (4) __________

Number of times random number generator cycles before simulation begins __________

Code for stat collection on monthly water yield

Skip collection (0) __________

Collect (1) __________

Code for stat collection on monthly sediment yield

Skip collection (0) __________

Collect (1) __________

*** SWRRB DATA ASSEMBLY FORM 2 ***

3. GENERAL DATA

Basin area (km^2) __________

Rainfall correction factor __________

Baseflow factor __________

Basin lag time (days) __________

Initial soil water storage -- fraction of field capacity __________

(entering 0. allows SWRRB to estimate it)

4. CENTROID COORDINATES OF SUB-BASINS

X: _______ _______ _______ _______ _______ Y: _______ _______ _______ _______ _______

5. GENERAL WEATHER DATA

If automatic entry is used, enter:

Weather station No. (Appendix II) __________ and skip to Section 10. If not, continue

TP-40 10-year frequency 0.5 hour rainfall (mm) __________

TP-40 10-year frequency 6.0 hour rainfall (mm) __________

Number of years for maximum monthly 0.5 hour rainfall record __________

Latitude of watershed (deg) __________

6A. MONTHLY TEMPERATURE DATA

Average monthly maximum air temperature (°C)

______ ______ ______ ______ ______ ______ ______ ______ ______ ______ ______ ______

Average monthly minimum air temperature (°C)

______ ______ ______ ______ ______ ______ ______ ______ ______ ______ ______ ______

Coefficient of variation for monthly temperature

______ ______ ______ ______ ______ ______ ______ ______ ______ ______ ______ ______

*** SWRRB DATA ASSEMBLY FORM 3 ***

6B. MONTHLY TEMPERATURE DATA FOR REMAINING SUB-BASINS

(SKIP UNLESS TEMPERATURE IS SIMULATED FOR EACH SUB-BASIN)

2 Maximum (°C)

______ ______ ______ ______ ______ ______ ______ ______ ______ ______ ______ ______

3

______ ______ ______ ______ ______ ______ ______ ______ ______ ______ ______ ______

4

______ ______ ______ ______ ______ ______ ______ ______ ______ ______ ______ ______

5

______ ______ ______ ______ ______ ______ ______ ______ ______ ______ ______ ______

2 Minimum (°C)

______ ______ ______ ______ ______ ______ ______ ______ ______ ______ ______ ______

3

______ ______ ______ ______ ______ ______ ______ ______ ______ ______ ______ ______

4

______ ______ ______ ______ ______ ______ ______ ______ ______ ______ ______ ______

5

______ ______ ______ ______ ______ ______ ______ ______ ______ ______ ______ ______

7. MONTHLY SOLAR RADIATION DATA Monthly average daily solar radiation (ly)

______ ______ ______ ______ ______ ______ ______ ______ ______ ______ ______ ______

8. MONTHLY RAINFALL DATA Monthly maximum 0.5 h rainfall (mm)

______ ______ ______ ______ ______ ______ ______ ______ ______ ______ ______ ______

9. PRECIPITATION GENERATION PARAMETER TABLE (Enter P(W/D) & P(W/W) or DAYPR)

P(W/D)

______ ______ ______ ______ ______ ______ ______ ______ ______ ______ ______ ______

P(W/W)

______ ______ ______ ______ ______ ______ ______ ______ ______ ______ ______ ______

Days Prec

______ ______ ______ ______ ______ ______ ______ ______ ______ ______ ______ ______

Mean

______ ______ ______ ______ ______ ______ ______ ______ ______ ______ ______ ______

St. dev.

______ ______ ______ ______ ______ ______ ______ ______ ______ ______ ______ ______

Skew coefficient

______ ______ ______ ______ ______ ______ ______ ______ ______ ______ ______ ______

*** SWRRB DATA ASSEMBLY FORM 4 ***

10. BASIN DATA Subbasin:	1	2	3	4	5	Entire basin
Fraction of basin in each subarea	____	____	____	____	____	
SCS runoff curve number - Condition II	____	____	____	____	____	
Soil albedo	____	____	____	____	____	
Water content of snow on ground at start of simulation (mm)	____	____	____	____	____	
Main channel length in each subbasin (km)	____	____	____	____	____	____
Average channel slope in each subbasin (m/m)	____	____	____	____	____	____
Average main channel width (m)	____	____	____	____	____	
Effective hydraulic conductivity of channel alluvium (mm/hr)	____	____	____	____	____	
Channel N value	____	____	____	____	____	____
Overland flow N value	____	____	____	____	____	____
Return flow travel time (days) (skip if unknown)	____	____	____	____	____	

Sediment concentration in return flow (ppm) ______ ______ ______ ______ ______

USLE erosion control practice factor P ______ ______ ______ ______ ______

Average slope length for each sub-basin (m) ______ ______ ______ ______ ______ ______

Average slope steepness for each
sub-basin (m/m) ______ ______ ______ ______ ______ ______

11. ROUTING DATA (If only one sub-basin is used, skip to Section 12)

(All routing data is from sub-basin to basin outlet)

Average channel width (m) ______ ______ ______ ______ ______

Average channel depth (m) ______ ______ ______ ______ ______

Channel slope (m/m) ______ ______ ______ ______ ______

Channel length (km) ______ ______ ______ ______ ______

Channel N value (mm/hr) ______ ______ ______ ______ ______

Effective hydraulic conductivity of channel
alluvium (mm/hr) ______ ______ ______ ______ ______

USLE soils K factor for channel ______ ______ ______ ______ ______

USLE C factor for channel ______ ______ ______ ______ ______

*** SWRRB DATA ASSEMBLY FORM 5 ***

12. POND DATA (If not used skip to Section 13)	Sub-basin: 1	2	3	4	5
Fraction of each sub-basin that flows into ponds	______	______	______	______	______
Total surface area of all ponds in each sub-basin (ha)	______	______	______	______	______
Runoff volume from pond catchment area to fill ponds (mm)	______	______	______	______	______
Initial pond volumes (mm)	______	______	______	______	______
Initial sediment concentration in ponds (ppm)	______	______	______	______	______
Normal sediment concentration in ponds (ppm)	______	______	______	______	______
Hydraulic conductivity of pond bottoms (mm/hr)	______	______	______	______	______

13. RESERVOIR DATA (If not used skip to Section 14)

Fraction of each sub-basin that flows in reservoirs	______	______	______	______	______
Total reservoir surface area at emergency spillway (ha)	______	______	______	______	______
Runoff volume from reservoir catchment area required to fill emergency spillway (mm)	______	______	______	______	______
Total reservoir surface area at principal spillway (ha)	______	______	______	______	______
Runoff required to fill to principal spillway (mm)	______	______	______	______	______
Initial reservoir volumes (mm)	______	______	______	______	______
Average principal spillway release rate ($m^3/s/km^2$)	______	______	______	______	______
Initial sediment concentration in reservoirs (ppm)	______	______	______	______	______
Normal sediment concentration in reservoirs (ppm)	______	______	______	______	______
Hydraulic conductivity of reservoir bottoms (mm/hr)	______	______	______	______	______

*** SWRRB DATA ASSEMBLY FORM 6 ***

14. SOILS DATA

Number of soils ____________

Soil Series name

#1 ____________ #2 ____________ #3 ____________ #4 ____________ #5 ____________

Sub-basin: 1 2 3 4 5

Soil Series for each sub-basin:

____________ ____________ ____________ ____________ ____________

No. layers for each soil series (if automatic entry is used, skip to Section 15)

____________ ____________ ____________ ____________ ____________

Soil Series name ____________________ K Factor __________ Pass #200 Sieve __________

Layer: 1 2 3 4 5 6 7 8 9 10

Depth to bottom layer (mm)

______ ______ ______ ______ ______ ______ ______ ______ ______ ______

Bulk density (T/m^3)

______ ______ ______ ______ ______ ______ ______ ______ ______ ______

Available water capacity (mm/mm)

______ ______ ______ ______ ______ ______ ______ ______ ______ ______

Saturated conductivity (mm/h)

______ ______ ______ ______ ______ ______ ______ ______ ______ ______

Clay content (%)

______ ______ ______ ______ ______ ______ ______ ______ ______ ______

Sediment size distribution (fraction-if unknown, skip to next soil)

Sand (.20 mm) Silt (.01 mm) Clay (.002 mm) Sm Ag (.03 mm) L Ag (.50 mm)

________ ________ ________ ________ ________

Maximum rooting depth (mm) ________

*** SWRRB DATA ASSEMBLY FORM 7 ***

Sub-basin ___________

15. CROP DATA

Planting date: Month ___________ Day ___________

Harvest date: Month ___________ Day ___________

(Use start and end of growing season for perennial vegetation)

Vegetation: Annual (1) ___________ Perennial (2) ___________

Tillage Operation:

Fall plow (1) ___________

Spring plow (2) ___________

Conservation tillage (3) ___________

Zero tillage (4) ___________

Average annual C-factor ___________

Maximum leaf area index ___________

Initial residue cover (kg/ha) ___________

16. IRRIGATION DATA

Are the crops irrigated Yes (1) ___________ No (0) ___________

Water stress when irrigation is applied ___________

Irrigation runoff ratio ___________

17. DAILY RAINFALL AND TEMPERATURE DATA

If rainfall is input, list daily rainfall (mm) for each sub-basin.

If temperature is input, list maximum and minimum temperature for each sub-basin in a separate file.

[Format (5X, 15F5.1)]

Appendix X

Output Variables

Table X.1. Definitions of output variables.

AIR	=	Average annual irrigation applied to each sub-basin (mm)
ATP	=	Sediment yield loss from ponds (t/ha)
ATR	=	Sediment yield loss from reservoirs (t/ha)
BST	=	Subsurface basin storage (mm)
CM	=	Average annual C-factor for each sub-basin
CS	=	Final sediment concentration in ponds (ppm)
CSR	=	Final sediment concentration in reservoirs (ppm)
D50	=	Median particle size of sediment (mm)
DFPQ	=	Pond water balance (mm)
DFPY	=	Pond sediment balance (t/ha)
DFRQ	=	Reservoir water balance (mm)
DFRY	=	Reservoir sediment balance (t/ha)
DMM	=	Average annual biomass production for each sub-basin (kg/ha)
FC	=	Available soil water content (mm)
IRN	=	Average number of irrigation applications per year for each sub-basin
K1-K9	=	Weather generator seeds
NPR	=	Number of peak flows
PHU	=	Potential heat units (C)
PL	=	Water yield loss from ponds (mm)
PR1	=	Mean peak flow/volume (m^3/s/mm)
PRB	=	Maximum peak flow rate (m^3/s)
PRM	=	Mean peak flow rate (m^3/s)
PSV	=	Standard deviation of peak flow rate (m^3/s)
PSZ	=	Sediment particle size distribution
RSDM	=	Average annual residue on ground surface for each sub-basin (kg/ha)
RZL	=	Minimum sub-basin soil water state for simulation (mm)
SBP	=	Average annual precipitation for sub-basins (mm)
SID	=	Final soil water content (mm)
SL	=	Slope length and steepness factor for MUSLE
SM2	=	Mean SCS curve number for simulation
SQ	=	Average annual surface runoff for sub-basins (mm)
SSB35	=	Composite soil water state (mm)
SSB35	=	Final composite soil water state for entire basin (mm)
SSBP	=	Standard deviation of rain storms (mm)
SSQ	=	Average annual subsurface flow for sub-basins (mm)
ST	=	Initial soil water content (mm)
ST	=	Final soil water state for each soil layer (mm)
SV	=	Final composite pond water volume (mm)
SW	=	Final soil water state (soil water minus 15-bar water) for each sub-basin (mm)
SWW	=	Initial composite soil water state (beginning of simulation) (mm)
SWW	=	Soil water balance at end of simulation (end of simulation) (mm)
SY	=	Average annual sediment yield for sub-basins (t/ha)
T1	=	Total porosity (mm/mm)
T2	=	Wilting point (1500 kPa) (mm/mm)
T3	=	Field capacity (33 kPa) (mm/mm)
TAV	=	Daily, average monthly, and average yearly temperature (C)

TC = Time of concentration for sub-basins and basin (h)
TL = Water yield loss from reservoirs (mm)
TOT = Final composite reservoir water volume (mm)
V = Final water volume in ponds (mm)
VB = Maximum SCS curve number during simulation
VL = Minimum SCS curve number during simulation
VR = Final water volume in reservoirs (mm)

The following variables are output on daily, monthly, yearly, and average annual basis. For example: SSB1 = daily precip, SMM1 = total monthly precip, SMY1 = total yearly precip, and SM1 = average annual precip. These values are weighed for the entire basin on the basis of sub-basin areas. Only SM is listed for convenience.

SM1 = Precipitation (mm)
SM3 = Surface runoff (mm)
SM4 = Subsurface flow (mm)
SM5 = Percolate out of soil profile (mm)
SM6 = Total water yield (mm)
SM7 = Evapotranspiration (mm)
SM10 = Solar radiation (ly)
SM12 = Sediment yield (t/ha)
SM36 = Snow fall (mm)
SM37 = Snow melt (mm)
SM38 = Channel transmission losses (mm)

The following variables are printed output only on an average annual basis.

SM11 = Total sub-basin sediment yield (t/ha)
SM13 = Sediment inflow to ponds (t/ha)
SM14 = Sediment outflow from ponds (t/ha)
SM16 = Sediment inflow to reservoirs (t/ha)
SM17 = Sediment outflow from reservoirs (t/ha)
SM19 = Pond evaporation (mm)
SM20 = Pond seepage (mm)
SM21 = Rainfall on ponds (mm)
SM22 = Water inflow to ponds (mm)
SM23 = Water outflow from ponds (mm)
SM24 = Reservoir evaporation (mm)
SM25 = Reservoir seepage (mm)
SM26 = Rainfall on reservoirs (mm)
SM27 = Water inflow to reservoirs (mm)
SM28 = Water outflow from reservoirs (mm)